SHAPED CRYSTAL GROWTH

FLUID MECHANICS AND ITS APPLICATIONS
Volume 20

Series Editor: **R. MOREAU**
MADYLAM
Ecole Nationale Supérieure d'Hydraulique de Grenoble
Boîte Postale 95
38402 Saint Martin d'Hères Cedex, France

Aims and Scope of the Series

The purpose of this series is to focus on subjects in which fluid mechanics plays a fundamental role.

As well as the more traditional applications of aeronautics, hydraulics, heat and mass transfer etc., books will be published dealing with topics which are currently in a state of rapid development, such as turbulence, suspensions and multiphase fluids, super and hypersonic flows and numerical modelling techniques.

It is a widely held view that it is the interdisciplinary subjects that will receive intense scientific attention, bringing them to the forefront of technological advancement. Fluids have the ability to transport matter and its properties as well as transmit force, therefore fluid mechanics is a subject that is particulary open to cross fertilisation with other sciences and disciplines of engineering. The subject of fluid mechanics will be highly relevant in domains such as chemical, metallurgical, biological and ecological engineering. This series is particularly open to such new multidisciplinary domains.

The median level of presentation is the first year graduate student. Some texts are monographs defining the current state of a field; others are accessible to final year undergraduates; but essentially the emphasis is on readability and clarity.

For a list of related mechanics titles, see final pages.

Shaped Crystal Growth

by

Y. A. Tatarchenko
Centre for Nuclear Energy,
Grenoble, France

KLUWER ACADEMIC PUBLISHERS
DORDRECHT / BOSTON / LONDON

Library of Congress Cataloging-in-Publication Data

```
Tatarchenko, V. A. (Vitalii Antonovich)
    [Ustoichivyi rost kristallov. English]
    Shaped crystal growth / by V.A. Tatarchenko.
       p.   cm. -- (Fluid mechanics and its applications ; v. 20)
    ISBN 0-7923-2419-6
    1. Crystal growth.   I. Title.   II. Series.
QD921.T3713   1994
548'.5--dc20                                          93-27851
```

ISBN 0-7923-2419-6

Published by Kluwer Academic Publishers,
P.O. Box 17, 3300 AA Dordrecht, The Netherlands.

Kluwer Academic Publishers incorporates
the publishing programmes of
D. Reidel, Martinus Nijhoff, Dr W. Junk and MTP Press.

Sold and distributed in the U.S.A. and Canada
by Kluwer Academic Publishers,
101 Philip Drive, Norwell, MA 02061, U.S.A.

In all other countries, sold and distributed
by Kluwer Academic Publishers Group,
P.O. Box 322, 3300 AH Dordrecht, The Netherlands.

Printed on acid-free paper

CONTENTS

Preface   ix

Introduction   1

I.1.   Capillary shaping techniques – CST   2
I.2.   The Czochralski technique – CzT   3
I.3.   The techniques of pulling from shaper – TPS   5
I.4.   The Verneuil technique – VT   9
I.5.   The Floating-zone technique – FZT   10
I.6.   Vapor whisker growth – VWG   11
I.7.   Cylindrical pore in a growing crystal – negative crystal   12
I.8.   The two-shaping element technique – TSET   14
I.9.   The variable shaping technique – VST   14
I.10. The local shaping technique – LST   18

**Chapter 1.   Dynamic Stability of Crystallization – The Basis of
                Shaped Crystal Growth**   19

1.1.   Formulation of the problem   19
1.2.   Melt meniscus shaping conditions   21
1.3.   Growth angle   26
1.4.   Mass-transfer processes on the crystal surface in the vicinity
       of the triple point   34
1.5.   Equation of the crystal dimension change rate   37
1.6.   Temperature distribution in the crystal-melt system   38
1.7.   Equation of the crystallization front displacement rate   41
1.8.   Convection influence on the liquid-phase heat-transfer processes   42

Chapter 2.   The Czochralski Technique   45

2.1.   Lyapunov set of equations   45
2.2.   Liquid meniscus height and shape (the capillary problem)   47
2.3.   The heat problem   60
2.4.   Crystallization process stability   67

Chapter 3.   Techniques of Pulling from Shaper   71

3.1.   Boundary conditions of the capillary problem   71
3.2.   Review of publications on capillary shaping in TPS   76
3.3.   Liquid meniscus shapes   78
3.4.   Capillary problem for systems with large Bond numbers   79
3.5.   Capillary problem for systems of large Bond numbers with an
       allowance for surface-tension gradients   95
3.6.   Capillary problem for systems of small Bond numbers   101
3.7.   Capillary problem with the gravity force allowed for   123
3.8.   Calculation of the transition of the wetting boundary condition to
       the catching one   25
3.9.   Influence of pressure and shaper to crystal clearance on the shape
       of profile curves   130
3.10. Approximate calculation of the melt-meniscus height   130
3.11. Limits of capillary equation applicability depending upon the
       Bond number value   133
3.12. Capillary coefficients A_{RR} and A_{Rh} 134
3.13. Capillary stability   141
3.14. The conditions for crystallization stability   142
3.15. Stationary values of R^0 and h^0 as simultaneous solutions of the
       capillary and heat problems   143
3.16. Growth of tubular crystals   145
3.17. Influence of melt pressure variations on crystallization stability   155
3.18. Selection of shaping conditions   159
3.19. Experimental test of some statements of the theory of capillary shaping 160

Chapter 4.   The Verneuil Technique   167

4.1.   Growth of cylinder- and plate-shaped crystals   167
4.2.   Stability analysis-based automation of the Verneuil technique   175
4.3.   Optimization of the crystallization process while growing crystals
       under unstable conditions   178
4.4.   Growth of tube-shaped crystals   179

Chapter 5.   The Floating-Zone Technique 187

5.1.   Problem formulation 187
5.2.   The heat conditions 188
5.3.   Mass balance 189
5.4.   The capillary coefficients 189
5.5.   Crystallization stability 193

Chapter 6.   Radial Instability of Vapor Whiskers 197

6.1.   Problem formulation 198
6.2.   Whisker growth 199
6.3.   Stability study 203
6.4.   Profile curves of initial whisker sections 205

Chapter 7.   Stability of Cylindrical Pore Growth 207

7.1.   Formulation of the problem 207
7.2.   The capillary coefficients A_{RR} and A_{RC} 207
7.3.   The concentration coefficients A_{CR} and A_{CC} 208
7.4.   Analysis of the process-stability conditions 209

Chapter 8.   Two Shaping Elements Technique 211

8.1.   Capillary shaping 211
8.2.   Hydrodynamics conditions 211
8.3.   Stability investigation 215
8.4.   Impurity distribution 215
8.5.   Application of carbon fiber-reinforced substrates 218
8.6.   Silicon as a component of the SOF material 219
8.7.   Prospects of carbon-silicon material application 220

Chapter 9.   Variable and Local Shaping Techniques 223

9.1.   Formulation of the problem 223
9.2.   Capillary coefficients A_{ll} and $A_{l\phi}$ 225
9.3.   Thermal coefficients $A_{\phi\phi}$ and $A_{\phi l}$ 225
9.4.   Analysis of stability 226

Chapter 10.   Crystal Cross-Section Shapes   227

10.1. Plate-shaped crystal pulling by TPS   227
10.2. Allowance for morphological instability of the crystallization front   231
10.3. Growth of circular cylinder-shaped crystals by TPS   233

Chapter 11.   Shaped Crystal Growth by TPS   237

11.1. Impurity distribution   237
11.2. Growth of silicon crystals   242
11.3. Growth of shaped LiNbO$_3$ crystals   252
11.4. Growth of shaped sapphire crystals by TPS   259

Chapter 12.   Crystallization under Microgravity Conditions   269

12.1. Simulation experiments   270
12.2. Crystallization of copper under short-time microgravity conditions   270
12.3. "Shape" experiment   273
12.4. "Ribbon" experiment   276
12.5. Conclusion   276

**Appendix. Calculation of the Growth Angle from the Shape of a
        Crystallization Drop**   279

References   283

PREFACE

The monograph "Shaped Crystal Growth" by V. A. Tatarchenko is the first systematic statement of the macroscopic crystallization theory. The theory is based on the stable-growth conception, which means that self-stabilization is present in the system, with growth parameter deviations occurring under the action of external perturbations attenuating with time. The crystallization rate is one of the parameters responsible for crystal defect formation. Steady-state crystal growth means that crystallization-rate internal stabilization is present, thus allowing more perfect crystals to grow. Most important is the fact that the crystal shape (an easily observed parameter) is one of the stable-growth characteristics when growing crystals without any contact with the crucible walls. This means that constant-cross-section crystal growth is to a certain extent evidence of crystallization process stability.

The principles of the stable crystal growth theory were developed by the author of the monograph in the early 1970s. Due to the efforts over the past 20 years of V. A. Tatarchenko, his disciples (V. A. Borodin, S. K. Brantov, E. A. Brener, G. I. Romanova, G. A. Satunkin *et al*) and his followers (B. L. Timan, O. V. Kolotiy *et al*) the theory has been completed, which is demonstrated by this monograph. The characteristic feature of the theory is its trend towards solving practical problems that occur in the process of crystal growth. Initially developed to account for the crystal shaping mechanisms of the Stepanov technique, the theory has been extended to the Czochralski, Verneuil and the floating-zone techniques. Calculation of steady-state conditions for growing tube-shaped corundum crystals by the Verneuil technique, and selection of control laws for the Verneuil technique are especially impressive results of practical application of the theory.

The stable growth theory is based on strong groundwork, i.e the Lyapunov stability theory. Only fundamental relations are used to derive the Lyapunov equations for crystal growth processes. This allows one to take theoretical recommendations without hesitation, many of them being experimentally proved. Practical steady-state crystal growth is also discussed in the monograph. It should be noted that not only conventional techniques of crystal growth but also original ones developed at the Institute of Solid-State Physics of the Russian Academy of Sciences with active participation of the author of the monograph are described here; they are the variational shaping technique and the two-shaping element technique.

No doubt, the success of stable growth theory is due to V. A. Tatarchenko, whose fundamental work was presented for publication in 1971 [1]. In 1974 it was presented at the International Conference of Crystal Growth in Japan [2], while the first work on stability in the EFG technique was published by Surek and Chalmers in 1975 [3].

I hope that specialists working in the field of solid-state physics and fluid mechanics find this book to be of interest.

This monograph is a translation of the Russian edition of the monograph "Stable Crystal Growth", published by Nauka publishing house, Moscow, in 1988 [4]. This edition has been extensively revised. The author added new data to chapters 7, 9 and paragraphs 11.3, 11.4, 12.1, 12.2, 12.4. The monograph is composed as follows.

The introduction presents a brief summary of the characteristics of the capillary shaping techniques which are further analyzed in terms of the crystallization process stability conditions. These techniques include the Czochralski, techniques of pulling from shaper (Stepanov, EFG, CAST, etc.), Verneuil, floating-zone, variational and local shaping, two-shaping element techniques as well as whisker growth according to the vapor-liquid-crystal (VLC) mechanism and cylindrical pore growth.

Chapter 1 offers a general approach to crystallization process stability analysis during capillary shaping. Each of the crystallization techniques is characterized by a finite number of degrees-of-freedom, independent variables. On the basis of general laws, i.e., the laws of conservation of energy, mass and the growth angle constancy condition specific for melt crystallization, a system of differential equations for these quantities is solved. The system is linearized and then analyzed for stability (the Lyapunov stability).

In the subsequent chapters, this procedure is consecutively applied to analyze all the crystallization techniques described in the introduction. A separate chapter deals with general problems of growing crystal shape stability. Experiments on growing profiled silicon, lithium niobate, and corundum single crystals that find wide engineering application as well as space crystal growth are also described separately.

The monograph represents a complete entity, which is why it is difficult to read chapters independently. The formulas and figures are numbered by chapters, while reference numbers are given for the whole monograph.

Prof. Yu. A. Osipian

INTRODUCTION

The development of many advanced fields in modern engineering is to a large extent governed by the success achieved in crystal growth. The development of the first solid-state lasers became possible only when perfect ruby crystals with uniform dopant distribution had been grown. Microminiaturization in electronics became widespread only after the improvement of dislocation-free semiconductor crystal structure since defects of micron-fraction size are comparable with integrated circuit elements. The above-mentioned examples are sufficient to prove the thesis. Naturally, engineering needs the development of stimulated crystallization.

Though works on vapor-phase crystallization have recently been rather intensive, the technology of commercial growth of bulk crystals is largely based on melt crystallization. Almost all melt-growth techniques have been used for many decades but during recent years they seem to have undergone a rebirth. Now perfect silicon and sapphire single crystals weighing tens of kilograms can be produced by melt crystallization. These achievements are accounted for not only by an improvement in equipment but also by further development of our understanding of the crystallization process.

However, modern engineering does not only need crystals of arbitrary shapes but also plate-, rod-, and tube-shaped crystals, i.e., crystals of shapes that allow one to use them as final products without additional machining. Therefore, crystals of specified sizes and shapes with controlled defect and impurity structures are required to be grown.

This problem appears to be solved by profiled-container crystallization as in the case of casting. However, this solution is not always acceptable. Container material needs to satisfy a certain set of requirements: it should neither react with the melt nor be wetted by it, it should be of a high-temperature and aggressive-medium resistant, etc. Even if all these requirements are satisfied, perfect-crystal growth is not secured and growing very thin plate-shaped crystals, to say nothing of more complicated shapes, excludes container application completely.

I.1. Capillary Shaping Technique – CST

Therefore, the techniques which allow the shaping of the lateral crystal surface without contact with the container walls rank high among the melt crystallization techniques, i.e., the Czochralski technique – CzT [5], the Stepanov technique – ST [6], edge defined film fed growth – EFG [7], capillary action shaping technique – CAST [8], the Verneuil technique – VT [9], the floating-zone technique – FZT [10] and various modifications thereof. Absence of contact between the crystallizing substance and the crucible walls in these techniques enables improvement of crystal structures, a decrease in mechanical stress levels and the crucible becomes unnecessary for the Verneuil and floating-zone techniques. For all these techniques the shapes and the dimensions of the crystals grown are controlled by the interface meniscus-shaping capillary forces and by the heat- and mass- exchange conditions in the crystal-melt system. Let the above-mentioned techniques be called capillary shaping techniques – CST.

Absence of rigid shaping and the advantages mentioned cause a number of problems when growing crystals of specified shapes and controlled cross-sections. Indeed, as the crystal is not restricted by the crucible walls, its cross-section depends upon the growing parameters. Any deviations of the growth rate, temperature conditions, etc. result in crystal cross-section changes, in pinch formation.

Experiments have demonstrated that an increased amount of inclusions and non-uniform impurity distributions can be observed at the pinch locations. It is not the pinches themselves that seem to cause defect formation, but a change in the crystal dimensions which indicates some deviations of the growth condition (mainly, the crystallization rate) from the steady-state ones. From this point of view, constant cross-section crystal growth is to a certain extent evidence of crystallization process stability. Therefore, the problem of crystal cross-section stabilization arises. For the capillary shaping techniques the stabilization problem proved to be solvable by investigating the dynamic crystallization-process stability and selecting crystallization schemes and conditions that will ensure growth of crystals of specified shapes and controlled cross-sections.

This analysis was carried out for the first time by the author of the present monograph for the Czochralski and the Stepanov techniques [1] in 1971 and later by Surek and Chalmers [3]. In [1] the author was able to show the role of capillary and heat effects in the process of constant cross-section crystal growth including the case of shaper employment. During the twenty years that have passed since first publication, dynamic stability of the Czochralski, Stepanov, floating zone and Verneuil techniques has been investigated in detail and the steady-state conditions and schemes for crystallization have been selected.

The present monograph is mainly based on the works published by the author and his collaborators, and publications by his colleagues (references will be provided). For this reason, the assistance given by V. A. Borodin, S. K. Brantov, E. A. Brener, G. I. Romanova and G. A. Satunkin and fruitful contacts with V. V. Voronkov, O. D. Kolotiy, B. L. Timan and A. A. Chernov that contributed greatly to an overall understanding of the phenomenon described are highly appreciated.

Recently, interest in the problem under consideration has increased due to the development of computer-aided crystal growth analysis [11, 12, 13]. The development of a computer-controlled system of crystal growth makes it necessary to construct a physical model of the crystallization process, which provide the steady-state growth conditions. In case the crystallization system is unstable by itself, the introduction of a controller will enable the achievement of stability of the complex: the crystallization system plus a controller. Some mathematical models of a controller for the Czochralski technique were developed on the basis of our physical model of melt crystallization process stability [14].

Works on stability of the process of growing crystals of controlled cross-section shapes are a direct continuation of earlier work on morphological crystal-melt interface stability. They essentially use the liquid-solid interface shape stability analysis technique based on the investigation of the heat field and impurity distribution character applied by Temkin [15], Voronkov [16] and Mullins and Sekerka [17] and supplemented in our case with the capillary effects and melt-crystal-gas triple point condition. Furthermore, these processes are correlated and the loss of crystallization front morphological stability can lead to crystal shape stability loss [18].

Brief characteristics of capillary shaping techniques follow.

I.2. The Czochralski Technique – CzT

Where the Czochralski technique [5] is used, crystals grow from the melt-free surface onto the seed pulled upright (Figure I.1.). This technique has been known of for more than 75 years but it has only found wide practical application because of the development of semiconductor engineering and now it is essentially the main technique used in semiconductor crystal growth.

Melt-pulling based crystal growth techniques (the Czochralski technique being the most striking example) are often identified, even by specialists, with the techniques of glass and plastic pulling. Thus, P. Yorgensen [19] asserts: "Crystal growth by pulling is identical to glass pulling. Analysis of the phenomena occurring in the process of pulling reveals their absolute identity in all the cases and they differ only in values of the quantities used. Sheet pulling from an aluminum bath does not raise more problems than glass tape manufacturing". Indeed, outwardly these processes are very much alike, however "the difference in values of the quantities used" points to the fact that pulling materials of crystal structures has very little in common with glass pulling.

Firstly, in the process of crystal growth, melt-crystal phase transition occurs and the heat of this transition evolves at the crystallization front. Therefore, crystallization should be carried out with a temperature gradient sufficient to remove the heat inside the crystal. It should be noted that for some IR-transparent materials the characteristic radiation corresponding to the crystallization heat can ensure crystallization heat removal from the phase-transition boundary [20, 21, 22]. Secondly, glass viscosity

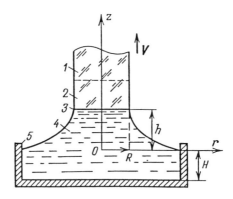

Fig. I.1. Crystal growth by the Czochralski technique: 1 – seed; 2 – crystal; 3 – crystallization front; 4 – melt; 5 – crucible; R – crystal radius; h – crystallization front height; V – crystal pulling rate; H – melt level in crucible; $z0r$ – coordinate system.

is higher and it gradually solidifies during pulling in the process of viscous flow, while liquid-state viscosity of the materials being crystallized is very low, therefore shaping during crystallization can be accomplished only by the capillary forces, there being no electrical, magnetic or other effects on the liquid. Since the capillary forces allow the formation of a meniscus with the height of the order of magnitude of the capillary constant (usually it is a few millimeters high), quite high stabilization of all the parameters is required during pulling. This is the main difficulty of melt-pulling crystallization techniques.

To formulate clearly the problems we encounter in the process of applying the melt pulling techniques, the Czochralski technique will be described below.

When a wettable body is immersed into the liquid, an equilibrium liquid column embracing the surface of the body immersed is formed. The column formation is caused by the capillary forces present. Such liquid configuration is usually called a meniscus. Note that for a body differing from a straight cylinder the height of liquid rising along its surface will be different in different sections. This height will depend on the wetting-angle value, body-surface curvature and slope in the area of its contact with the liquid (Figure I.2).

Let the temperature of the meniscus horizontal section be T_0, the temperature of liquid crystallization. So, above the plane of this section the liquid transforms into a solid phase (Figure I.1). Now we will set the liquid phase into upward motion with the constant rate, V, keeping the position of the phase-transition plane invariable by selection of the heat conditions. When the motion starts, the crystallized portion of the meniscus will continuously form a solid upward or downward tapering body. In a particular case when a line tangent to the liquid meniscus surface makes a specific angle (angle of growth) with the vertical, the crystallized portion of the meniscus will have the form of a cylindrical body whose cross-section will coincide with the

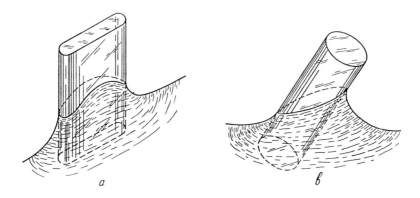

Fig. I.2. Meniscus shapes when a wettable body is immersed into a liquid: a straight elliptic cylinder-shaped body (a); an oblique circular cylinder-shaped body (b).

meniscus section formed by the crystallization plane (Figure I.1).

Thus, the initial body (called the seed hereafter) serves to form a meniscus which later on determines the form of the product crystallized, the phase transition surface position being fixed. Then a specimen of specified shape is relatively easy to pull just by selecting proper seed and meniscus and by positioning the phase transition plane into a certain place. However, it is difficult to obtain desired shapes of crystallized products in such a way. It is easy to understand why. In the process of pulling, the liquid surface and phase transition surface positions are subjected to perturbations that affect the shape of the product being pulled. This in its turn results in changes of the position of each meniscus point, with a new equilibrium surface of the liquid being formed. These perturbations accumulate with time and the shape of the profile being pulled is distorted.

The idea of steady-state crystal growth is to provide negative feedback in the crystallization system by utilizing the inherent reserves of the system. In this case the perturbations occurring in the system are suppressed, which ensures crystallization process stability and leads to the possibility of growing crystals of constant cross-section shapes.

I.3. Techniques of Pulling from Shaper – TPS

TPS differ from the Czochralski technique by the presence of a shaping device on the melt surface (Figure I.3). It should be noted that the method of inserting holes into plates placed on the melt surface for shaping melt-pulled crystals was primarily developed by Gomperz [23] and Kapitza [24] but it was due to methodical works carried out by Stepanov and his collaborators that the technique was further refined and it is now widely used for the crystallization of various materials. Conferences

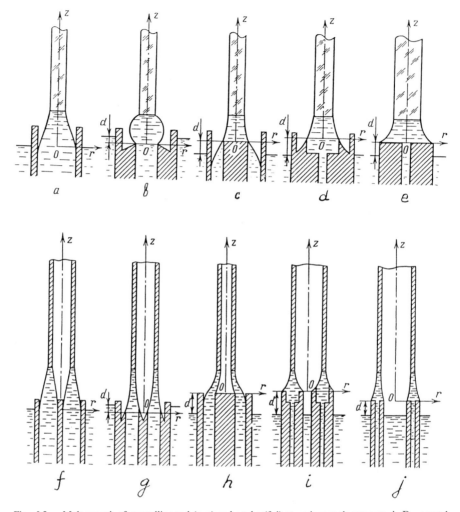

Fig. I.3.   Melt growth of crystalline rod (a–e) and a tube (f–j) at various melt pressure d, R – crystal radius, h – crystallization front height.

were held in Russia [25] and it was decided that all techniques of crystal growth from shaper were to be named after Stepanov.

In recent years, developers of other techniques, such as the EFG (edge defined film fed growth [7, 26]), the CAST (capillary action shaping technique [8]) and some others [27, 28] have become opposed to the use of the Stepanov technique in the literature. To solve this problem, we suggest combining all these techniques by using a common title, TPS – techniques of pulling from shaper. In fact, the main idea of all these techniques is to limit the area and the value of liquid-free surface perturbations occurring in Czochralski crystal growth during melt crystal-pulling, which must contribute to controlled shaping. As was shown in our analysis [29, 30, 31], only two possibilities can be used for shaping under these conditions – the edges or walls of a shaper. From a mathematical point of view, this corresponds to two boundary conditions of the capillary problem – a fixation of line on the surface of meniscus or a fixation of angle between the melt and the melt of shaper [29, 30, 31]. All the techniques mentioned use this principle of shaping, but each one is characterized by a design feature – using external (Figure I.3e) and internal (Figure I.3b, d) free edges, surfaces (Figure I.3a) and a combination thereof (Figure I.3c) for solid bodies in contact with the melt. Such devices will be referred to as shapers. For products of complicated shapes (Figure I.3f–j) the shaper represents a combination of elements shown in Figure I.3a–e. Some positive or negative pressure, d, can be applied to the melt. This can be achieved by controlling the position of the shaper-free edges relative to the liquid free surface level (Figure I.3).

Based on this description, a conclusion can be drawn that the dimensions and shapes of the specimens being pulled by TPS depend upon the following factors: (i) the shaper geometry; (ii) the pressure of feeding the melt to the shaper; (iii) the crystallization surface position and shape; (iv) the seed shape. The seed shape is only important for stationary pulling, in this case its cross-section should coincide with the desired product cross-section. Frequently, especially when complicated profiles are grown, a wettable rod can be used as a seed, its main function being to raise the melt column above the shaper level. Then, the pulling process is carried out under unstationary conditions by lowering the crystallization surface, which then enhances the shaper dependence of the crystal cross-section. With such an approach applied to the pulling process, the dimensions and the shape of the crystal grown are determined by the above-mentioned factors and by the pulling rate-to-crystallization front displacement rate ratio [32].

As follows from the description of the process, the shaper in TPS does not serve as a gauged orifice that strictly determines the dimensions and the shape of the crystal being pulled. For this reason, it is not advisable to use the term "die" instead of the term "shaper" as is often done; for the extrusion die strictly determines the profile cross-section. In TPS, crystallization takes place without contact between the specimen being crystallized and the solid body. It is this condition that enables the development of thin walls and crystal surfaces of good quality.

To grow a crystal of specified cross-section and dimensions by TPS, a corresponding shaper design should be selected and it should be ensured that the crystallization surface is kept at the level required during the pulling process. While doing this, one should take into consideration that stabilization of the melt temperature and the pulling rate are insufficient for keeping the crystallization surface position unchanged as the crystal length increase and melt volume decrease, change the heat conditions within the growth area.

The shaper is one of the main elements of the crystallization heat zone in TPS. To demonstrate the problems that are to be solved in the process of designing it, we

Fig. I.4.   Growth of crystal of complicated shape with the difference in melt rise height at the sections of various curvature compensated by the shaper.

will analyze a very simple example, melt pulling of a circular cylindrical rod. Even when one shaping scheme is used this rod can be grown using shapers of various dimensions and various pressures (Figure I.3b, d). That is why the criteria allowing proper selection must be developed. The problem becomes still more complicated if crystal shape differs from the circular one. In this case, the shaper can compensate for the difference in the melt-rise height caused by varying crystal surface curvature to provide flat crystallization front during pulling (Figure I.4).

The above description does not demonstrate completely the shaper role in TPS. Furthermore, it does not explain the differences between TPS and the Czochralski technique. Indeed, in TPS, shaper walls can be moved off the crystal (Figure I.3). In the Czochralski technique, crystal size can be commensurable with the crucible size (Figure I.1). In which of the cases can the crucible wall effect on shaping be ignored? Where does the Czochralski technique end and the TPS begin?

The results of the crystallization process stability study answered these questions [1]. The shaper proved not only to limit the area of melt perturbations as supposed but in many cases it stabilizes the crystallization process. Therefore, considerable attention will be given to the crystallization process stability investigation with various schemes of shaping applied.

Profiles of complicated-shape cross-sections can be grown by TPS. Figure I.5 shows an aluminum alloy heat-exchanger and other profiles from aluminum, grown by us in Stepanov's laboratory. The one on the left consists of an external shell, an internal eight-tube panel and six cross pieces (cross pieces are removed). The accuracy of maintaining profile dimensions can be judged by the fact that: the heat-exchanger is more than one meter long and it was assembled without any mechanical treatment of the profiles.

Fig. I.5. Aluminum heat-exchangers and other profiles from aluminum alloys melt-grown by TPS in Stepanov's lab.

I.4. The Verneuil Technique – VT

Figure I.6 illustrates Verneuil crystal growth [9]. Fine powder of the material a crystal is grown from is fed into a burner flame. The powder melts partially or completely, forming a melted layer on the surface of the seed crystal. The crystal grows in the process of consecutive melt crystallization as a result of sinking the crystal into the colder zone.

In the classic version of VT, gaseous hydrogen and oxygen are fed into the burner. Later on, versions using plasma, radiation heating, etc. appeared. However, they did not change the principal scheme.

At the beginning of the century, this technique was exclusively used to grow precious crystals from refractory oxides. Considerable changes to the equipment used in the technique resulted from its application to growing ruby single crystals for laser elements. However, no ideas about the optimal temperature distribution for this

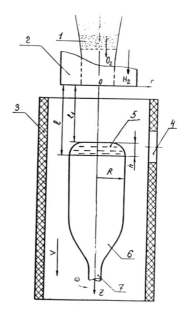

Fig. I.6. Cylindrical crystal growth by the Verneuil technique: 1 – stock bin; 2 – burner; 3 – furnace thermal insulation; 4 – peephole; 5 – melt layer; 6 – crystal; 7 – seed; R – crystal radius; l_1 – melt surface position relative to the burner; $h = l - l_1$ – melt meniscus height; $z0r$ –coordinate system; V – displacement rate; ω – crystal rotation rate.

technique could be found in the literature, only very general calculations according to simplified schemes were made [33].

Attention in some experiments was diverted to the fact that growth of ruby crystals of small diameters (< 4 mm) was stable, they had smooth surfaces and cylindrical shapes. This phenomenon was explained in our works on the crystallization process stability of Verneuil growth [34]. Recommendations on heat-zone design securing process stability follow from these works. Practical application of these recommendations has led to growing one side-sealed tubular crystals [35] (Figure I.7).

I.5. The Floating-Zone Technique – FZT

A polycrystalline rod with a melt zone formed on it by a concentrated heat source is an initial material for the floating-zone technique. The zone moves up and down the rod forming a single crystal (Figure I.8). In case the position of the melting rod is fixed relative to the crystal being grown, the diameter of the growing crystal is on average equal to the diameter of the initial rod. The growing crystal diameter can be

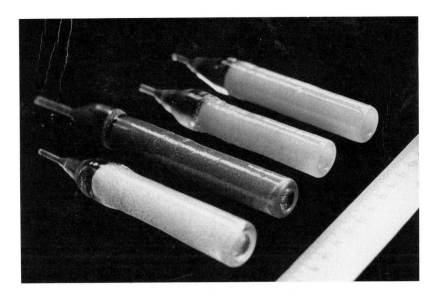

Fig. I.7. Monocrystalline sapphire and ruby tubes grown by the Verneuil technique under stable conditions.

changed by increasing or decreasing the zone dimensions ($V > < V_m$).

A radio-frequency induction heater is most frequently used as the heat source. The technique is widely used in industry for growing silicon single crystals.

From the point of view of the crystallization process stability analysis, the floating-zone technique exhibits an important feature. Not only crystallization interface but also melting interface are exhibited here. This made it necessary for additional studies of the boundary conditions of the capillary shaping problem at the melting interface [36].

I.6. Vapor Whisker Growth – VWG

In the 1960s, whiskers attracted everybody's attention as their strength proved to be close to the theoretical one. Despite the fact that their unique properties have not yet been fully exploited, they still excite both scientific and practical interest. The main whisker growth technique is vapor growth.Their growth by the vapor-liquid-crystal (VLC) transition is now considered to be proven [37, 38] (Figure I.9). The vapor-phase substance is absorbed by the melt at the crystal end face and then it crystallizes according to the scheme resembling the Verneuil technique. The only difference is that flux growth takes place as the melt is enriched with a great amount of impurities used for controlled whisker nucleation on the substrata. This whisker crystallization scheme can be regarded as a capillary shaping technique and its stability can be

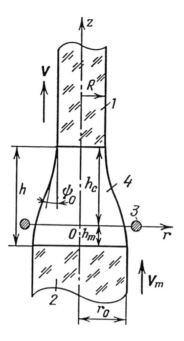

Fig. I.8.   Cylindrical crystal growth by the floating zone technique: 1 – growing crystal with the radius R; 2 – feeding rod with the radius r_0; 3 – heater; 4 – melted zone; h_c, h_m – positions of crystallization front and melting front relative to the heater, respectively; ψ_0 – growth angle; V – rate of growing crystal displacement relative to the induction heater; V_m – the same for melting rod.

studied according to a general procedure.

I.7. Cylindrical Pores in a Growing Crystal – Negative Crystals

Foreign particles trapped by growing crystals, including gas pores, represent a very common defect in crystals. Such imperfections are currently attracting increasing attention. The nature of the porosity of crystals depends on many factors such as the material of the crystals, the tendency of the melt to dissociate and to interact with the material of the container, the morphology of the crystallization front, the hydrodynamics of various flows in the system, the technique of growth, the rate of pulling, and a number of other physical and technological factors. Pore formation in growing crystals is associated with very different causes in the literature, which indicates their diversity in particular situations.

    In this chapter, the dynamic stability approach is used to investigate the conditions of elongated cylindrical pore (negative crystals) formation and steady growth (Figure I.10) [39].

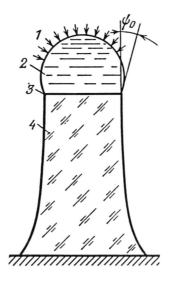

Fig. I.9.   Whisker growth according to the VLC-mechanism: 1 – vapor; 2 – flux; 3 – crystallization front; 4 – crystal; ψ_0 – growth angle.

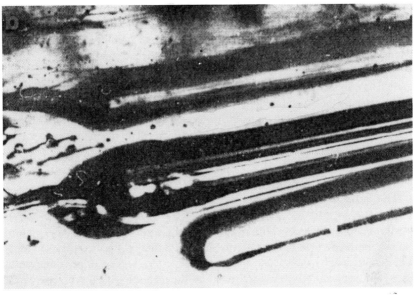

Fig. I.10.   Optical pattern of gas inclusions in the sapphire crystal [39].

I.8. The Two Shaping Element Technique

The manufacture of cheap silicon sheets is one of the most important problems of modern semiconductor metallurgy. Its solution would to a considerable extent determine further development of solar power engineering and micro electronics.

The processes of epitaxial growth of crystalline silicon layers on substratum require as a rule expensive single crystal substratums and equipment and allow one to solve only certain problems of microelectronic technology.

The great potential of amorphous silicon growth on cheap polycrystalline or amorphous substratums cannot be realized because of the low efficiency and stability of solar cells based on it and unsuitability of this material for other purposes.

Silicon profiling by TPS allows the manufacture of thin-walled crystals suitable for solar cell production. Investigation of the capillary shaping during such crystal growth reveals that the growth meniscus length is of the order of a hundred micrometers. Because of this fact, the operator would need to be highly skilled, furthermore, this process is difficult to automate.

A new melt crystallization technique allowing more than an order of magnitude increase of the growth meniscus length has been developed [28]. This is achieved by using two shaping elements, a melt wettable support substrata and a feeder, a device for controlled growth meniscus formation, placed between the main melt volume and the substrata and connected with the main melt volume by capillary channels. A horizontal version of the technique is shown in Figure I.11.

The feeder provides control of a number of crystallization process parameters. When comparing this technique with the technique of melt crystal growth on substratums known of up to now [40], we can notice that the difference between them is very similar to that between TPS and the Czochralski techniques. The feeder can be regarded as a shaper. Unlike the shaper, the feeder is not the only shaping element as the geometry of the crystal being grown is also determined by the substrata. When graphite cloth is used as a substrata, the technique is usually called the SOC technique ("silicon-on-cloth") [41, 42]. We will consider the SOC technique as a particular case of the two shaping element (USE) technique.

I.9. The Variable Shaping Technique – VST

Growth of profiled crystals with variable cross-section shapes is a very important problem. In the process of crystal growth by TPS, the profile cross-section shape is mainly controlled by the shaper design, and its cross- section dimensions can vary only within very narrow limits restricted by the meniscus-existence zone, with the melt meniscus catching on the shaper free edge. Therefore, the first attempt to widen the limits of varying pulled profile cross-section dimensions in the process of growth was to displace movable shaper elements [43].

The second technique of varying profile dimension, e.g. tube inner diameters, is based on melt spreading from a circular capillary channel over the shaper solid

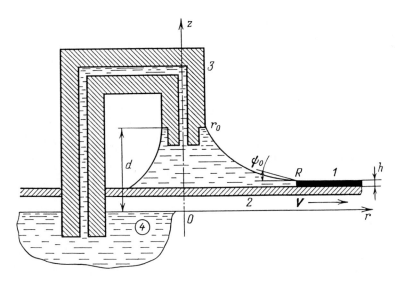

Fig. I.11.   Crystal growth by the two-shaping element technique – TSET: 1 – tape; 2 – substrate; 3 – feeder; 4 – melt volume; V_0 – substrate displacement velocity; h - crystallized layer thickness; ψ_0 – growth angle; r_0 – feeder free-edge position relative to the feeder axis; d – melt pressure in the meniscus; R – crystallization front position relative to the feeder axis.

surface, with proper melt temperature and pulling rate control provided [26].

When the techniques of relative displacement of the movable part of the shaping device and control of melt spreading over the shaper solid surface were applied, a number of disadvantages emerged. Even in the case of precise manufacture of shaper movable elements catching occurred.

Practically, one shaper could not be used for more then two growing procedures because of deformation caused by temperature changes. The second technique is difficult to reproduce when tubes with diameters exceeding 15 mm are grown because of process instability caused by the absence of the inner meniscus catching on the shaper-free edge.

A technique of profiled crystal growth called the variable shaping technique was developed [44, 45, 46]. It possesses none of the disadvantages mentioned above and allows changes within wide limits to both the dimensions and the shapes of the crystal cross-sections during the growth process. The variable shaping technique is based on controlling the melt mass flow towards the crystallization interface when passing to a new specified cross-section shape. In this case, different shaping elements of the same shaper, its free edges or walls, operate by turns. Periodic changes of the tubes' diameter and attachment and break of side ribs of the growing tube at a specified moment present no difficulties.

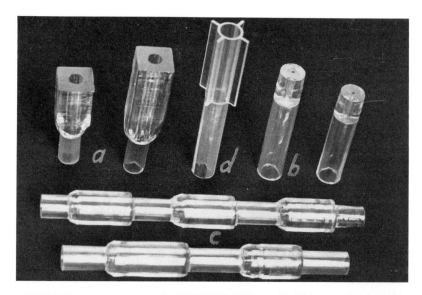

Fig. I.12.   Monocrystalline sapphire profiles with variable cross-section shapes grown by VST.

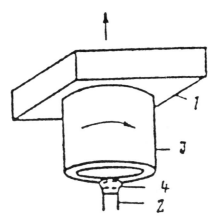

Fig. I.13. Schematic drawing of LST realization tube-growth [47]: 1 – seed; 2 – shaper; 3 – grown crystal; 4 – meniscus.

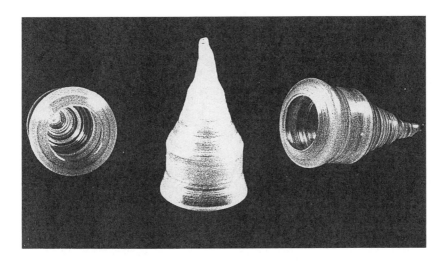

Fig. I.14. Some complex sapphire products grown by LST [49].

Transition from a rectangular cross-section profile to a tube and back is also easy to perform. Corundum crystals of complicated shapes and crucibles of various diameters, 16, 25 and 30 mm, with wall thickness of 2.5 mm have been produced (Figure I.12).

The variable shaping technique allows improvement to the group growth techniques, i.e., growing a group of profiles using one seed crystal. A profile with the cross-section containing all the elements of the future profile group and cross pieces between them is formed in the process of growth. When growth is completed, change

for a new cross-section shape without cross pieces is carried out and the group of profiles grows independently. The variable shaping technique can be applied to growing
profiled crystals of various materials.

I.10. The Local Shaping Technique – LST

The local shaping technique is one of the many versions of TPS. In contrast to
the conventional TPS, in this process, the shaper forms only an element of the
predetermined cross-section of the crystal being pulled, while the whole cross-section
is obtained by means of crystal rotation (Figure I.13). This technique was applied
to grow lithium fluoride tubes with the external diameter of 20 mm and 4 mm thick
walls [47], sapphire disks 60 mm in diameter [48] and complex sapphire products in
the form of hollow cones and cone-shaped tubes with continuously alternating side
surface [49, 50] (Figure I.14).

Chapter 1

DYNAMIC STABILITY OF CRYSTALLIZATION – THE BASIS OF SHAPED CRYSTAL GROWTH

1.1. Formulation of the Problem

Various crystallization schemes and conditions will be analyzed from the point of view of growing crystals of specified cross-sections under capillary-shaping conditions [1, 2, 31, 51–59].

All the crystallization techniques will be characterized by a finite number, n, of the main variables, degrees-of-freedom, X_i, that can arbitrarily vary in the process of crystallization.

Thus, for the Czochralski technique they will include the crystal dimensions R, and the crystallization front position relative to the melt-free surface, h (Figure I.1). For TPS they include the crystal dimensions, R, and the crystallization front position relative to the shaper free edges, h, (Figure I.3), i.e., these techniques have two degrees-of-freedom ($n = 2$). For more exact analysis, these techniques can be represented as systems with three degrees-of-freedom where the third degree-of-freedom is the pressure, d, under which the melt is fed to the shaper for TPS (Figure I.3) and the melt level, H, in the crucible for the CzT (Figure I.1).

For the Verneuil technique, the melt-free surface position relative to the burner level, l_1, i.e. the melted layer thickness, h, is added to the crystal dimensions, R, and the crystallization front position, l (Figure I.6). This technique has three degrees-of-freedom ($n = 3$).

The floating-zone technique has four degrees-of-freedom ($n = 4$); they are the crystal dimensions, R, the melted zone volume, W, and the positions of crystallization surface, h_c, and melting surface, h_m, relative to the heater (Figure I.8).

In order to grow a crystal of constant cross-section, stabilization of the above-mentioned parameters should be provided during the crystallization process. Here, stabilization of the crystallization front position means that the crystallization rate should be equal to the crystal pulling rate at any moment, i.e., no random changes of the crystallization rate can be observed. Naturally, such stabilization can be provided by means of automatic control systems. At the first stage of the analysis our objective is to ensure stabilization of the parameters specified using the internal functions of the system and only where it is not possible to find the law of control that can ensure crystallization process stability.

Growth of a crystal of constant cross-section is a particular case of the problem of growing a crystal of controlled cross-section. The latter problem can be solved by changing $(n - 1)$ of the parameters X_i (naturally, excluding the crystal dimensions) and the crystallization conditions (pulling rate, melt temperature, etc.) taking into consideration the functional dependence of all the parameters X_i and the crystallization conditions found beforehand.

To find the explicit function, a set of simultaneous equations should be solved. The set of equations should include: 1) the melt or solution flow equation (the Navier-Stokes equation) with the boundary conditions on the meniscus-free surface (the Laplace capillary equation); 2) the continuity equation (the law of crystallizing substance mass conservation); 3) the heat-transfer equation for the liquid and the solid phases with the equation of heat balance at the crystallization front and at the melting front as the boundary conditions (the law of energy conservation); 4) the diffusion equation (impurity mass conservation), and 5) employment of the growth angle constancy condition specific for melt crystallization. The set of equations mentioned above is general for all the crystallization techniques under consideration, while the specific features of each of the techniques and schemes are characterized by the set of boundary conditions and concrete values of the parameters included in the equations. The approach proposed here is valid if heat and mass transfer is the limiting stage of the process rather than crystallization kinetics, which is usually done for real systems and pulling rates. Then such kinetic effects as the dependence of crystallization front supercooling on the crystallization rate or the impurity effect on the crystallization temperature can be taken into account within the model developed here.

The number of equations used should be equal to the number of degrees-of-freedom, n, in the crystallization scheme under consideration, multiple use of each of the equations mentioned, e.g., the growth angle constancy conditions, being possible for investigation of inner and outer surface shaping when tubular crystals are grown.

As will be shown below, the following set of differential equations is obtained for the set of X_i:

$$\frac{dX_i}{dt} = f_i\left(X_1, X_2, \ldots, X_n, \frac{dX_1}{dt}, \frac{dX_2}{dt}, \ldots, \frac{dX_n}{dt}, t, C\right) \tag{1.1}$$

$$i = 1, 2, \ldots, n.$$

Here t denotes the time, n is the number of the unknowns that depends on the crystallization technique and the growing crystal cross-section, C denotes a set of controllable crystallization parameters, the thermophysical and other constants of the substance being crystallized.

As was mentioned above, the stationary values of the variables X_i^0 will be especially interesting for us, i.e., growing crystals of constant cross-sections with stationary crystallization-front positions and other parameters of the system. These values satisfy Equation (1.1) with zero left side

$$f_i(X_1^0, X_2^0, \ldots, X_n^0, t, C) = 0 \tag{1.2}$$

Only those stable solutions of (1.2) can be physically realized. According to Lyapunov [60] the solutions of (1.1) are stable if they are stable for the linearized set of equations:

$$\delta \dot{X}_i = \sum_{k=1}^{n} \frac{\partial f_i}{\partial X_k} \delta X_k \tag{1.3}$$

Here $\delta X_k = X_k - X_k^0, \delta X_i = X_i - X_i^0, \delta \dot{X}_i = \frac{d}{dt}(\delta X_i)$, all partial derivatives are taken with $X_k = X_k^0$. The stability of (1.3) in its turn is observed when all the roots S in the characteristic equation

$$\det\left(\frac{\partial f_i}{\partial X_k} - S\delta_{ik}\right) = 0 \tag{1.4}$$

have negative real components, δ_{ik} is the Kronecher delta [60]. This equilibrium will be unstable if (1.4) has at least one root with a positive real component, and unless an imaginary number can be found among the roots, additional study including an allowance for the non-linear terms in (1.3) is required.

Calculation of the time-dependent nonstationary functions f_i requires the solution of (1.1) and is usually rather difficult. These difficulties can be avoided by using the quasi-stationary approach. For example, Mullins and Sekerka [17] applied this approach to the temperature and impurity distribution problem while studying the morphological stability of the crystallization-front shape, which led to the results that agreed well with the experiment. However, in each particular case the quasi-stationary approach is to be found.

A number of constraints imposed on the systems and perturbations occurring in the course of the Lyapunov stability study should be noted. Stability is examined over an infinitely long period of time. In this case, the perturbations are considered to be small and are only imposed on the initial conditions, i.e., the same forces and energy sources affect the system after the perturbations as before.

1.2. Melt Meniscus Shaping Conditions

For the capillary shaping techniques the crystal cross-section is determined by the melt meniscus section formed by the crystallization surface. The melt meniscus shape can be calculated on the basis of the Navier-Stokes equation, the Laplace capillary equation being the free-surface boundary condition.

The full-scale solution of this problem offers considerable mathematical difficulties. Therefore, to simplify the problem formulation, the contributions of various factors of meniscus shaping should be estimated: inertial forces associated with the

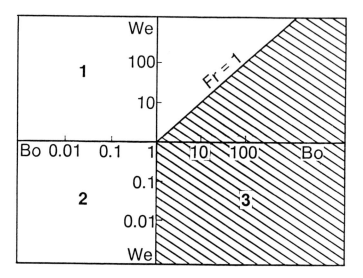

Fig. 1.1.   Inertial (1), gravitational (2) and capillary (3) force effects on melt column shaping [62] (We, Bo, Fr are characteristics Weber, Bond and Froude numbers, respectively).

melt flow, capillary forces, gravity forces, viscous and thermocapillary forces [61]. The influence of the latter two factors will be discussed in the corresponding sections. The relative effect of the first three factors can be estimated by means of dimensionless numbers: the Weber number $We = \rho_L V^2 L/\gamma_{LG}$ characterizing comparative action of the inertial and capillary forces, the Froude number, $Fr = V/(gL)^{1/2}$, characterizing comparative action of the inertial and gravity forces; the Bond number, $Bo = \rho_L g L^2/\gamma_{LG}$, characterizing comparative action of the gravity and capillary forces. Here ρ_L denotes liquid density, L denotes the liquid meniscus dimensions, V is the liquid flow rate, γ_{LG} denotes the liquid surface tension coefficient, g relates to the gravity acceleration. When the Weber and Froude numbers are small, the melt flow can be neglected. The Bond number defines the region of capillary or gravity force predominance (Figure 1.1). The effect of the inertial force as compared with the gravity and capillary forces proves to be negligible, for all the crystallization rates used – as a first approximation, liquid flow rate is considered to be equal to the crystallization rate. Indeed, if we assume that the linear dimensions of the meniscus lie within the range of $10^{-3} - 10^{-2} m, \rho_L \approx 10^{-3} - 10^{-4} kg m^{-3}, \gamma_{LG} \approx 1 N m^{-1}$ can be examined in the hydrostatic approximation up to the fluid speed of $(0.1\text{-}1.0)$ m s^{-1}. Convective flows whose rates can substantially exceed the crystallization rate can occur in a liquid column in addition to the flow associated with crystallization. These flow effects on meniscus shaping and on liquid-phase heat-transfer will be discussed below.

1.2.1. *The Meniscus Surface Equation*

In the hydrostatic approximation the equilibrium shape of the liquid surface is described by the Laplace capillary equation [63]

$$\frac{\gamma_{LG}}{R_1(M)} + \frac{\gamma_{LG}}{R_2(M)} + \rho_L gw = const. \tag{1.5}$$

Here $R_1(M)$ and $R_2(M)$ denote the main radii of liquid surface curvature at the point M of the liquid surface. The radii of curvature are considered to be positive if the positive direction of the normal is inside the liquid, and negative if it is outside the liquid. In this, the positive direction of the normal towards a curve on the meniscus surface coincides with the direction from the center of curvature towards a given point of it. The w-axis is directed vertically upwards. The value of *const* depends upon w-coordinate origin selection and is equal to the pressure on the liquid in the plane $w = 0$. In particular, if the w-coordinate origin coincides with the plane of the liquid surface, *const* = 0.

However, to calculate the meniscus shape for the capillary-shaping techniques, it is convenient to employ the Laplace equation in its explicit differential form. To obtain this form of the equation, following [63] the variational problem of the total free energy minimum of a liquid column will be solved. For our system the total free energy is composed of the surface free energy and the gravity field energy: for the uvw-coordinate system it gives:

$$\gamma_{LG} \int_S dS + \int_w dW = \min \tag{1.6}$$

Here S denotes surface, W denotes melted-zone volume. The meniscus surface equation in the form $w = f(u, v)$ is sought, then (1.6) will have the following form:

$$\int\int_S \left(\gamma_{LG} \left[1 + \left(\frac{\partial w}{\partial u} \right)^2 + \left(\frac{\partial w}{\partial v} \right)^2 \right]^{1/2} - \frac{1}{2}\rho_L gw^2 \right) dudv = \min \tag{1.7}$$

The minimum should be calculated with the additional condition of liquid volume constancy taken into account:

$$\int_W dW = const \tag{1.8}$$

The Euler equation is an indispensable condition of functional extremum existence (1.7) and in accordance with [64, 65] it has the following form:

$$\text{div}(\hat{H} \text{ grad } w) - \frac{\rho_L g}{\gamma_{LG}} w = const' \tag{1.9}$$

Here,

$$\hat{H} = \left[1 + \left(\frac{\partial w}{\partial u}\right)^2 + \left(\frac{\partial w}{\partial v}\right)^2\right]^{-1/2} \tag{1.10}$$

const' is opposite in sign to *const* in (1.5). Let *const'* = $-P$.

Solving the boundary problem for (1.9) presents some difficulties. Therefore our study will be restricted to considering meniscus possessing axial symmetry. Such meniscus are obtained during melt pulling of straight circular cylinder-shaped crystals. We will find the equation of such meniscus surface, by introducing the cylindrical coordinates $w = w, u = r^* \cos \varphi, v = r^* \sin \varphi$ and taking into account the meniscus shape independence on the azimuthal angle, φ, Equation (1.10) will be rearranged into the following form:

$$\frac{w''}{(1 + w'^2)^{3/2}} + \frac{w'}{r^*(1 + w^2)^{1/2}} + \frac{1}{\gamma_{LG}}(P - \rho_L g w) = 0. \tag{1.11}$$

It is important to note that the transition from (1.5) to (1.11) is associated with substantial constraints: the meniscus surface is assumed to have one-valued projections onto the $0uv$-plane at each point, while for two-valued meniscus each branch is described by Equation (1.11) with different signs before the last term depending upon the sign of w'. Here the prime denotes differentiation with respect to r^*, the plus sign in the equation corresponds to the meniscus part with $w' < 0$, the minus sign corresponds to the meniscus part with $w' > 0$.

The problem of liquid meniscus shape calculation for an axially symmetric meniscus is reduced to finding the shape of a profile curve $w = f(r^*)$, the liquid surface meniscus being obtained by rotating this curve about the w-axis. Now we will introduce the capillary constant, a, and pass to the dimensionless coordinates and parameters

$$(2\gamma_{LG}/\rho_L\gamma)^{1/2} = a, w/a = z, r^*/a = r, Pa/(2\gamma_{LG}) = d.$$

This transition means that the capillary constant serves as a linear dimension unit, and the weight of a liquid meniscus one capillary constant high corresponds to the pressure equal to one. The approach allows application of the calculated results to any substance and various gravity accelerations, with only the scale changing.

Then (1.11) takes the form:

$$z''r + z'(1 + z'^2) \pm 2(d - z)(1 + z'^2)^{3/2}r = 0. \tag{1.12}$$

For large Bond numbers ($Bo > 1$; Figure 1.1) when gravity prevails (this condition corresponds to growing crystals with diameters of the order of ten capillary constants) (1.12) can be simplified:

$$z'' \pm 2(d - z)(1 + z'^2)^{3/2}r = 0. \tag{1.13}$$

For small Bond numbers ($Bo < 1$, Figure 1.1) when capillarity prevails (this condition corresponds to growing crystals with a diameter smaller than the capillary constant and can be easily satisfied by decreasing the gravity when the capillary constant is high), Equation (1.12) can also be simplified

$$z''r + z'(1 + z'^2) \pm 2d(1 + z'^2)^{3/2}r = 0. \tag{1.14}$$

In some cases, it will be more convenient for us to examine the Laplace capillary equation not in the form of $z = f(r)$ but in the form of $r = f(z)$. This transformation is usually used if the function $z = f(r)$ is not single-valued, while transition to the function $r = f(z)$ removes this multi-valuedness.

In this case (1.12) has the following form:

$$r''r + (1 + r'^2) + 2(d - z)(1 + r'^2)^{3/2}r = 0. \tag{1.15}$$

Accordingly, (1.13) and (1.14) can be transformed.

1.2.2. *Static Stability of the Meniscus*

It should be mentioned that the Laplace capillary equations follow from the indispensable condition of functional extremum existence (1.7). Naturally, not all the meniscus that represent solutions of the Laplace capillary equation will be stable, i.e., will be physically realized. This meniscus stability (the static one) should be distinguished from the dynamic stability of the crystallization process discussed above. For statistically stable meniscus, not only indispensable but also sufficient conditions of functional extremum (1.7) should be satisfied.

These are the following conditions [66]:

Firstly, the solution of the Jacobi differential equation should have no conjugate points. For a general case the Jacobi differential equation has the following form:

$$\frac{d}{dx}\left[\frac{\partial^2 f}{(\partial y')^2}\eta' + \frac{\partial^2 f}{\partial y \partial y'}\eta\right] - \left[\frac{\partial^2 f}{\partial y^2}\eta + \frac{\partial^2 f}{\partial y \partial y'}\eta'\right] = 0. \tag{1.16}$$

Here $f(x, y, y')$ denotes an argument of the functional under investigation and $y(x)$ is the Euler equation solution. Usually, (1.16) is solved as a boundary problem with the boundary conditions $\eta(x_0) = 0, \eta'(x_0) = 1$. If this equation takes on the zero value even at the points different from x_0, these points are referred to as conjugate points relative to x_0.

Secondly, the Legendre condition should be satisfied:

$$\text{if } \frac{\partial^2 f}{(\partial y')^2} > 0 \text{ the minimum solution is obtained} \tag{1.17}$$

$$\text{if } \frac{\partial^2 f}{(\partial y')^2} < 0 \text{ the maximum solution is obtained.} \tag{1.18}$$

The Jacobi equation and the Legendre conditions in their explicit forms will be written out for those specific cases for which meniscus static stability will be studied.

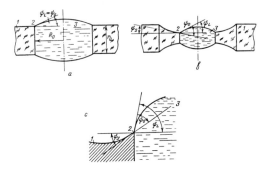

Fig. 1.2.   Calculation of the growth angle ψ_0 from the crystallized melted disk shape: disk thickness is constant, $\psi_L = \psi_0(a)$; decreasing disk thickness, $\psi_L < \psi_0(b)$; increasing disk thickness, $\psi_L > \psi_0(c)$: 1 – crystal; 2 – crystallization front; 3 – melt.

1.2.3. *Boundary Conditions of the Capillary Problem*

As the Laplace capillary equation is a second-order differential equation, formulation of the boundary problem for melt column shape calculation requires assignment of two boundary conditions. These boundary conditions are determined by the structural features of each specific capillary-shaping scheme and will be analyzed in detail below. In this case, one of the crystal-melt interface conditions is common for all the melt crystallization techniques. This condition follows from the condition of growth angle constancy.

1.3. Growth Angle

Let the angle ψ_0 (Figure 1.2) made by the line tangent to the meniscus and the lateral surface of the growing crystal be called the *growth angle*. The growth angle should not be confused with the *wetting angle*. The wetting angle characterizes particular equilibrium relative to the liquid movement along a solid body and is not directly associated with crystallization. The case of thermo-dynamically positioning equilibrium position of a melt drop on the crystal surface is an exception (see below, Figure 1.7).

    In the first works investigating the capillary-shaping conditions in the Czochralski and TPS a crystal of constant cross-section was considered to grow where the growth angle ψ_0 is equal to zero [67, 68]. Judging by a purely geometric diagram of liquid-solid phase conjugation, such an assumption is quite natural. However, investigation of the crystal growth process showed that the condition $\psi_0 = 0$ is not satisfied while growing crystals of constant cross-sections.

1.3.1. *Direct Measurements*

Direct measurements of this angle using photographs of melt columns in the process of Czochralski growth were made in [69, 70].

In [69], the growth angle for germanium and silicon crystallization was examined both by direct measurements using meniscus photographs and by comparing their heights obtained experimentally and those calculated by the formula from [71]. This formula comprises the liquid-solid phase conjugation angle as a parameter. In [69], special attention was paid to crystal diameter constancy in the process of growth angle measurements. For silicon the growth angle obtained was equal to $\approx 15^0$.

It should be noted that this work revealed that the melt-rise height is independent of the crystallization rate and the crystal rotation rate and is determined by the growing crystal diameter alone. This experimental fact confirms our statement made in the above paragraph of the present chapter concerning the validity of applying the Laplace capillary equation use instead of the Navier-Stokes equation for meniscus shape description during real-rate growth and rotation of crystals.

High-speed photography application [69] demonstrated that although ψ_0 is on average equal to 15^0, considerable fluctuations of its value can be observed in the process of growth. As shown in [70], the growth angle of germanium also differs from zero. However, the spread in ψ_0-values proved to be considerable: $\psi_0 \approx 20^0 - 30^0$, owing to the unsatisfied stationary crystal-pulling condition.

In [72] ψ_0 is identified with the angle of crystal wetting by the melt. In the case of silicon crystallization by the Czochralski technique the value of this angle was determined by the melt-rise height along the crystal facets making various angles with the vertical and by the melt drop shapes at the crystallization front. Unfortunately, the details of the ψ_0 calculation technique by melt drop shapes are not cited in this work. The growth angle value for silicon was obtained: $\psi_0 = (21 \mp 2)^0$. Note that Voronkov [73] gives the value of 35^0 for the angle of silicon surface wetting by its melt, and in [74] data on melt wetting for various germanium facets are presented: 30^0 for $\{111\}$, 17^0 for $\{110\}$, 9^0 for $\{100\}$.

In [75], the growth angle was determined in the process of copper crystallization by the Czochralski technique by direct measurements using photographs: $\psi_0 = (-0.6 \mp 0.4)^\circ$.

1.3.2. *Calculation From the Crystallized Disk Shape*

The technique from [3] is based on determining the crystallized melted disk shape. In such a way important errors caused by inconstancy of the growing crystal diameter and by subjectivity of identifying the liquid-solid phase contact-point position on the photographs, which is characteristic of the above-described techniques, can be avoided. An experimental diagram is presented in Figure 1.2. A melted zone was formed on a flat silicon or germanium washer of specified orientation and thickness by an electron beam. A specified amount of the material was added to this zone. Allowing for the initial substance mass within the zone, the zone volume was

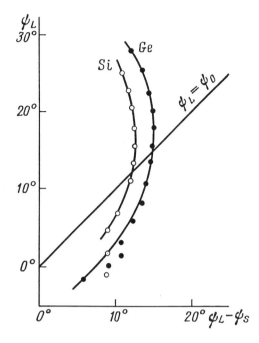

Fig. 1.3.   The relative angle value, $\psi_L - \psi_S$, as function of the angle, ψ_L (symbols see in Figure 1.2).

calculated. The meniscus was assumed to have the shape of a part of a sphere in all cases as the initial zone dimensions R_0 and r_0 are small and the gravity effect can be ignored. Therefore, ψ_L, the angle made by the line tangent to the melt in the process of its crystallization with the abscissa axis, can be easily calculated from the washer thickness and zone volume known. The growth angle $\psi_0 = \psi_L - \psi_S$ where ψ_0 is made by the line tangent to the crystallized zone with the abscissa axis. The crystallized-zone profile was obtained using its cross-section photograph. That allowed the calculation of the crystallized material mass and the slope of the line tangent to the zone profile curve at any point, i.e., to find the angles ψ_L, ψ_S. The authors believe that most valid values of the growth angles $\psi_0 = 11^0 \mp 1^0$ for silicon and $\psi_0 = 13^0 \mp 1^0$ for germanium) can be found when $\psi_S = 0$ (Figure 1.3). Calculated for the converging ($\psi_S < 0$) and the diverging ($\psi_S > 0$) parts of the zone, the growth angles differ from each other, the growth angle being smaller for the converging part and larger for the diverging one. Far from the extreme point $\psi_S = 0$, the difference is large and close to it the spread in values does not exceed $\mp 1^\circ$.

1.3.3. *Calculation From the Crystallized Drop Shape [76]*

With the crystal detaching itself from the melt, a drop is formed from the melt film trapped by the crystal at the crystallization interface. Melt film contraction into a drop is caused by incomplete melt wetting of the crystalline surface. With the position that can be characterized by the wetting angle θ achieved, the drop is in the state of mechanical equilibrium with the crystal. Then the crystallization process develops, which leads to cone-shaped crystal formation.

In our experiments [76] we grew and studied crystallized drops of silicon, germanium and indium antimonide obtained with a crystal detaching itself from the melt in the process of pulling by the Czochralski technique (Figure 1.4). X-ray topography of cross-section laps confirmed monocrystallinity of the drops crystallized.

A diagram illustrating the calculations is given in Figure 1.5. The angle $\psi_0 = \psi_L - \psi_S$ (Figures 1.5a, b) at any profile point can be easily found if the equation of the generator of a body of revolution resulting from melt drop crystallization is known. The angle ψ_L can be found by restoring the initial shape of the liquid segment using its known volume recalculated from the crystallized drop volume. The procedure used to calculate the specified angles is given in the Appendix.

Figure 1.6 shows the relationships $\psi_L(r)$, $\psi_S(r)$ and $\psi_0(r)$ for some drops. At the initial stage of crystallization, the calculated value of the growth angle is closest to the real one as the crystallization front keeps its initial shape. The values of $\psi_0 = 25^0 \mp 1^0$ for indium antimonide, $\psi_0 = 11^0 \mp 1^0$ for silicon and $\psi_0 = 12^0 \mp 1^0$ for germanium were obtained. The error was computed using the rms-error formulae. The growth angles calculated in such a way for Si and Ge agree with the values given in [3].

Some increase of the calculated ψ_S-values with drop solidification (Figure 1.6) can be associated with crystallization front curving. The front becomes convex and, as e.g., for InSb $\rho_S/\rho_L < 1$, the angle $\psi_S(r)$ proves to be larger than for the initially flat crystallization front.

The procedure of growth angle calculation discussed here allows the use of standard growth equipment used for Czochralski growth and requires no special equipment.

1.3.4. *Thermodynamic Approach*

The experimental data presented allow us to state that with the proof achieved in the experiment ψ_0 is a quantity constant for the given material. It should not be calculated from the geometric scheme as in [67, 68] but on the basis of examining the physical nature of the processes occurring in the vicinity of the crystal-melt-gas three-phase line.

Theoretically it was shown in [77–81]. We believe that a pattern of crystal lateral surface shaping can be constructed based on the data obtained in these works.

In 1960, Bolling and Tiller [82] showed that the crystallization front should be curved in the vicinity of the three-phase line (Figure 1.7a). The dimensions of the

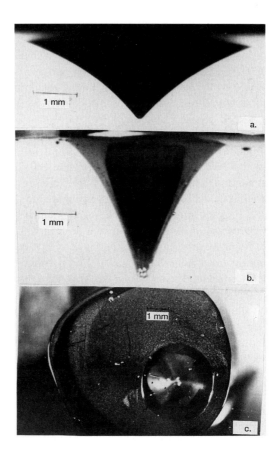

Fig. 1.4.   Photographs of detachment drops at the crystallization front: indium antimonide (a, c – top view), germanium (b).

curved area and the radius of its curvature for practical used growth parameters are equal to a few micrometers and depend upon the temperature gradient in the crystal. Here it is assumed that, firstly, the crystal and the melt are in thermodynamic equilibrium at the interface; secondly, the interface is isotropic and possesses the specific surface energy γ_{SL}; thirdly, the temperature change in the vicinity of the interface is linear along the z-coordinate.

If a crystal of the radius R is in thermodynamic equilibrium with the melt, its melting temperature differs from that of a flat crystal by the value of ΔT (the Gibbs-Thomson effect) [83]:

$$\Delta T = \gamma_{SL}/(\Delta S^* R) \qquad (1.19)$$

Here ΔS^* denotes melting entropy. Now we will examine a macroscopically flat

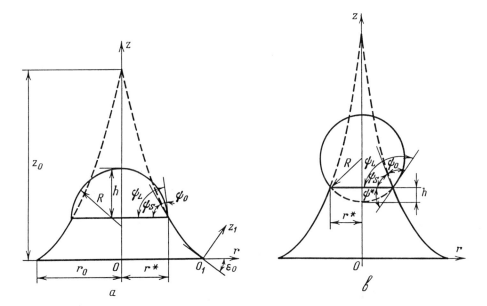

Fig. 1.5. Calculation of the growth angle, ψ_0, from the shape of crystallized melt drops: $\psi_L < \pi/2$ (a); $\psi_L > \pi/2$ (b).

crystallization surface in the coordinate system given in Figure 1.7a. For simplicity, we will regard the diagram as a two-dimensional one with the assumption that the crystallization surface shape is constant in the third dimension. Then from Equation (1.19) we have:

$$T(\infty, 0) - T(r, z) = (\gamma_{SL}/\Delta S^*)[-z''(1 + z'^2)^{-3/2}] \tag{1.20}$$

Here $T(\infty, 0)$ denotes the melting temperature for a flat surface. Taking into account that the temperature within the crystal is a linear function of z : $T(r, z) = G_S z$ and using the notation $K = (\gamma_{SL}/G_S \Delta S^*)$ from (1.20) we obtain:

$$K^2 z'' - z(1 + z'^2)^{3/2} = 0 \tag{1.21}$$

Equation (1.21) is an analog of the Laplace capillary equation in the form of (1.13) with $d = 0$, where K is used as the capillary constant. For the boundary conditions $z' \mid_{z=0} = 0$, $z \mid_{r=0} = \infty$, the equation of the solid-liquid interface shape is obtained in the form of an elementary function set:

$$r = K \ln \left[\frac{2K + (4K^2 - z^2)^{1/2}}{z} \right] - (4K^2 - z^2)^{1/2} + 0.53K \tag{1.22}$$

The liquid-solid phase conjugation angle (Figure 1.7a) is the boundary condition for the given model. It identifies the position of the liquid-solid phase contact point and,

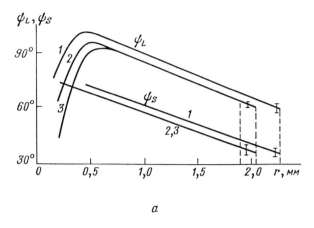

a

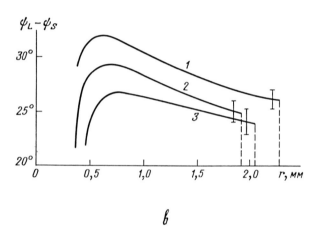

b

Fig. 1.6. ψ_L and ψ_S (a); $\psi_L - \psi_S$ (b) versus r for three indium anti-monide drops.

hence, the supercooled-liquid volume. The problem of the growth angle nature is not discussed here. At the same time the crystallization front curvature in the vicinity of the triple point mentioned here for the first time is an important element of the current schemes. In particular, since the curved-area dimensions are of the order of magnitude of K, for the crystal temperature gradient values used in the experiment's estimation of this area dimensions gives the value of the order of a few micrometers.

In [84], the relationship of the angles at the crystal-melt-gas triple point is calculated from heat-flow balance at the point mentioned. As follows from this work, the ψ_0-value should be determined by melt- and crystal-heat exchange with the environment and depends upon the angle the crystallization front makes with the crystal surface. However, this inference is drawn ignoring the crystallization front curvature in the vicinity of the triple point discussed above and furthermore, the limiting tran-

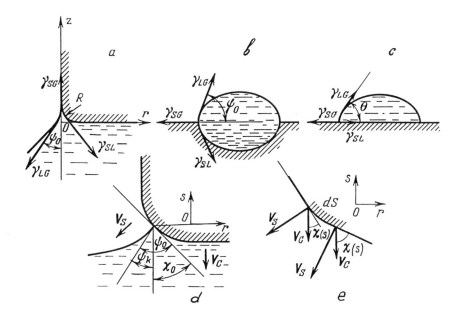

Fig. 1.7. Conjugation of three phases: solid, liquid and gaseous ones: in the process of crystal growth under near-equilibrium conditions (a); for thermodynamic equilibrium of the melt drop on the crystal surface (b); for thermodynamically nonequilibrium drop position on the crystal surface (c); with an allowance for kinetic effects during crystal growth (d, e).

sition to the triple point is used for temperature gradient calculation. This transition does not prove its value as the triple point is a singularity of the heat problem in the crystal-melt system.

In [85], the crystallization front angle relations are calculated from the surface tension force equilibrium at the crystal-melt, crystal-gas, melt-gas interfaces. Just the absence of equilibrium of the above-mentioned forces and hence systematic melt pulling-off account for step formation on the growing crystal surface. However, force equilibrium proves to be insufficient for calculation of the growth-angle values.

The growth-angle expression was first derived in [77] from the thermodynamic equilibrium condition at the three-phase interface line. Two conditions were used for selection of the interface displacement direction: firstly, an increment in the system free energy should be equal to zero for the three-phase interface to displace in the direction of the existing crystal-surface extension as this direction is an equilibrium displacement direction; secondly, an increment in the system free energy with the interfaces displacing in this direction should have an azimuthal-angle minimum. Hence,

$$\psi_0 = \arccos[(\gamma_{LG}^2 + \gamma_{SG}^2 - \gamma_{SL}^2)/2\gamma_{LG}\gamma_{SG}] \tag{1.23}$$

Here $\gamma_{LG}, \gamma_{SG}, \gamma_{SL}$ denote the free energies at the liquid-gas, solid-gas, solid-liquid interfaces, respectively.

In [81], a different approach to the equilibrium problem at the three-phase interface was developed. The authors allow for the free surface energy anisotropy at the crystal-melt and crystal-gas interfaces using the thermodynamic equilibrium condition along the interface of the existing phases obtained by Herring [86]:

$$\sum_i (\gamma_i r_i + \partial \gamma_i / \partial r_i) = 0 \qquad (1.24)$$

Here r_i denotes a unit vector normal to the line of interface intersection and lying in the ith plane of the interface. In fact, (1.24) signifies equality of forces and momenta of forces applied to each of the boundaries at the interface.

The expression for ψ_0 obtained in [81] differs from (1.23) in the correction factor, including the derivatives of the interface free surface energy. Hence, the correction factor results from the anisotropy and for an isotropic case is equal to zero. According to the authors' estimation the correction factor is small. This approach confirms the validity of (1.23) for the isotropic model and makes it possible to derive the growth angle expression with anisotropy being allowed for.

All the results of the present paragraph, excluding the data of [84], fit in with a single scheme if we assume that both the growth angle and the angle of crystal wetting by the melt are determined from the condition of triple-point thermodynamic equilibrium. It can be easily understood from Figure 1.7, where the crystal-melt boundary angles for the cases of crystal pulling (Figure 1.7a) with both equilibrium (Figure 1.7b) and non-equilibrium but mechanically balanced (Figure 1.7c) drop positions on the smooth crystal surface are shown. The spread in experimental data when determining the growth angle from the wetting angle should be explained by the deviations of the experimental conditions from the equilibrium ones.

This situation should be taken into account for the direct angle measurement techniques using crystal-pulling melt-column photographs. Here, small deviation from the thermodynamic equilibrium can lead to slight crystal melting. But as will be shown in the paragraph below, the kinetic effects in the vicinity of the triple point can significantly influence the experimental results during melting as the liquid-solid phase conjugation angle can be arbitrary in this case. Therefore, systematic slight crystal melting can account for the wide-spread in the growth angle values observed using high-speed photography [69].

1.4. Mass-Transfer Processes on the Crystal Surface in the Vicinity of the Triple Point

1.4.1. *Crystallization [80]*

As follows from the previous paragraph, the newly-formed crystal surface and the melt surface form a specific angle, ψ_0, whose value is determined from the thermodynamic equilibrium conditions. However, the newly-formed flat crystal portion is not in thermodynamic equilibrium with the melt. In this case, the difference in chemical potentials of the melt $\hat{\mu}_L$ and the flat crystal $\hat{\mu}_{S0}$ is equal to $\hat{\mu}_L - \hat{\mu}_{S0} = q\Delta T/T_0$, where q denotes the melting heat per atom, ΔT stands for melt supercooling and T_0 is the melting temperature. Owing to this difference in chemical potentials, substance migration from the melt onto the crystal surface will occur causing it to curve. The chemical potential, $\hat{\mu}_S$, of the isotropic crystal curved surface is equal to

$$\hat{\mu}_S = \hat{\mu}_L - q\Delta T/T_0 + \gamma_{SG}\zeta/N \qquad (1.25)$$

Here ζ denotes the crystal surface curvature, N denotes the number of atoms in a unit crystal volume. The migration will cease when the chemical potentials of the crystal and the melt are equalized. It means that at this moment the crystal surface curvature in the vicinity of the three-phase line is equal to

$$\zeta_e = \mathcal{L}\Delta T/(T_0\gamma_{SG}) \qquad (1.26)$$

Here $\mathcal{L} = qN$ relates to the melting heat of a unit crystal volume. Since far from the three-phase line the curvature is practically equal to zero, a curvature gradient, i.e., a chemical potential gradient that results in mass transfer, occurs along the crystal surface:

$$J = \lambda d\hat{\mu}_S/ds \qquad (1.27)$$

Here J denotes the substance flow, λ is the surface mass-transfer coefficient, $d\hat{\mu}_S/ds$ denotes the chemical potential gradient along the crystal surface S, ds is a crystal surface element, s relates to the distance from the three-phase line along the crystal surface (Figure 1.7e). From the condition of mass balance the relationship between the speed, V_S, along the normal towards the outer surface and the diffusion flow follows:

$$NV_S = -dJ/ds \qquad 1.28)$$

On the other side, V_S and the crystal growth rate V_C (V_C direction coincides with the direction of the crystal flat portion) are interrelated by the relationship (Figure 1.7e):

$$V_S = V_C \sin\chi(s) \qquad (1.29)$$

Here $\chi(s)$ is the angle between an element of the curved surface and the vector V_C. We substitute (1.29) and (1.27) into (1.28) and express $\hat{\mu}_S$ from (1.25). Allowing for $\zeta = d\chi/ds$, the following equation for $\chi(s)$ is obtained:

$$d^3\chi/ds^3 = -(N^2V_C/\lambda\gamma_{SG})\sin\chi \qquad (1.30)$$

Assuming that χ is small enough and thus $\sin\chi = \chi$, the solution of Equation (1.30) can be written down in the following form:

$$\chi(s) = \chi_0\exp(-Bs), \quad \text{where } B = (N^2V_C/\lambda\gamma_{SG})^{1/3} \qquad (1.31)$$

Using (1.26) as the boundary condition $d\chi/ds \mid_{s=0}= \zeta_e$, χ_0 can be defined:

$$\chi_0 = (\mathcal{L}\Delta T/T_0)(\lambda/V_C)^{1/3}(\gamma_{SG} N)^{-2/3} = k_\psi V_C^{-1/3}\Delta T \tag{1.32}$$

Finally, for ψ_k we obtain (Figure 1.7):

$$\psi_k = \psi_0 - \chi_0 = \psi_0 - k_\psi V_C^{-1/3}\Delta T \tag{1.33}$$

Thus, if crystallization is referred to as a kinetic process, the growth angle dependence on the three-phase line displacement rate V_C and the crystallization front supercooling ΔT occurs. The value of crystallization front supercooling is strictly dependent upon the growth mechanism. Thus, according to the estimations presented in [79] for silicon, the value of ΔT is equal to 3.7^0C when growing dislocation-free crystals if the crystallization front coincides with the singular facet and layer-by-layer growth takes place; $\Delta T \approx 0.3^0C$ when dislocations are formed in the crystal and $\Delta T \approx 0.05^0C$ during normal non-singular surface growth. In this case the kinetic correction to the growth angle in (1.33) changes from $\sim 20^0$ for dislocation-free crystals when the singular face is exposed at the crystallization front to $\sim 0.3^0$ for the normal mechanism.

In the first case the growth angle should be negative, which is a well-known experimental fact of dislocation-free silicon growth. In the case of normal growth, the kinetic correction is smaller than the accuracy of growth angle measurements in the experiments described above. It is therefore not surprising that the growth angle dependence on the growth rate and supercooling value was not detected experimentally.

As was noted above, we will primarily be discussing mass-transfer process-limited growth below, therefore we will regard the growth angle as the constant of the substance $\psi_k = \psi_0$, excluding the cases specified.

1.4.2. Melting [36]

As is known from experiments, in contrast to crystallization in the case of crystal melting the angle between the crystal surface and the melt can assume any value. This difference strikingly reveals itself in the floating-zone technique where crystallization takes place at one end of the melted zone and melting takes place at the other end of it. As the processes of crystallization and melting differ only in small deviations from the thermodynamic equilibrium, it is natural to explain the great difference between them by the kinetic effects.

During melting the melt is overheated, the rates of all the mass-transfer processes reverse, the sign of the right side of Equation (1.30) changes and the function:

$$\chi(s) = \exp(-1/2Bs)[\chi_0 \cos \sqrt{3}/2Bs) + b^* \sin \sqrt{3}/2Bs)] \tag{1.34}$$

will be the solution of this equation. Here b^* is the second integration constant. All other notations are the same as in the previous paragraphs. Then, in the case of melting, in contrast to crystallization, the boundary condition (1.26) is already insufficient to define the function $\chi(s)$ containing two integration constants and therefore the angle χ_0 is nondetermined. In terms of the character of the crystal-melt boundary condition, this fact is predominant since only melt fixation by the crystal free edge occurs during melting. As noted in the previous paragraph, the curvature radius and the curved-zone dimension are equal to 1 micrometer, i.e., the free edge can be considered as being sharp enough. Hence, in accordance with the terminology introduced below the catching boundary condition is satisfied at the crystal-melt boundary during melting. In this case, the characteristic time for attaining equilibrium is as follows:

$$\tau \sim (\lambda \gamma_{SG}/N^2 V_C^4)^{1/3} \tag{1.35}$$

For silicon when $\lambda \approx 10^{31} J^{-1} s^{-1}, \gamma_{SG} \sim 1 N m^{-1}, N \sim 5 \cdot 10^{28} m^{-3}, V_C \sim 10^{-5} m s^{-1}$, τ is of the order of 10^{-2}s. As will be shown below, it is substantially less than the time of the transient processes used in the crystallization stability analysis.

1.5. The Equation of Crystal Dimension Change Rate

Proceeding from the condition of growth angle constancy, we can obtain an equation for the crystal characteristic dimension change rate, R, from the geometric diagram given in Figure 1.8 [31, 51]. The vector **R** lies within the diagram plane and represents the radius for a straight circular cylinder-shaped crystal. For a plate it is its half-thickness. Now we will introduce the angles made by the line tangent to the meniscus on the three-phase line with the horizontal, α_0, and with the vertical, ψ (the crystal grows in the vertical direction Figure 1.8). When the angles α_0 and ψ add up to $\pi/2$, a crystal of constant cross-section R°, grows (Figures 1.8a, d) and in this case the angle ψ is equal to the growth angle ψ_0. Let α_e denote the values of the angle α_0 corresponding to this stationary-growth process: $\alpha_e = \pi/2 - \psi_0$. If $\alpha_0 \neq \alpha_e$, the crystal changes its dimension $\delta R = R - R^\circ$ in accordance with the law (Figures 1.8b, c, e, f):

$$\delta \dot{R} = V_C \text{tg}(\alpha - \alpha_0) = -V_C \text{tg}(\delta \alpha_0) \tag{1.36}$$

Here V_C denotes the crystallization rate. In this case the angle of crystal tapering at any moment is equal to $\delta \alpha_0 = \alpha_0 - \alpha_e$ and the crystallization rate V_C is equal to the difference in rates between pulling and front displacement:

$$V_C = V - dh/dt \tag{1.37}$$

As we have agreed to obtain the qualities characterizing Equation (1.3) near the stationary state, the deviations of $\delta \alpha_0$ and dh/dt can be regarded as negligible. Hence, by expanding $\text{tg}(\delta \alpha_0)$ into a Taylor series in the vicinity of the stationary point and using only the terms of the first-order infinitesimal we have:

$$\delta \dot{R} = -(V - dh/dt)\text{tg}(\delta \alpha_0) \approx -V \delta \alpha_0 \tag{1.38}$$

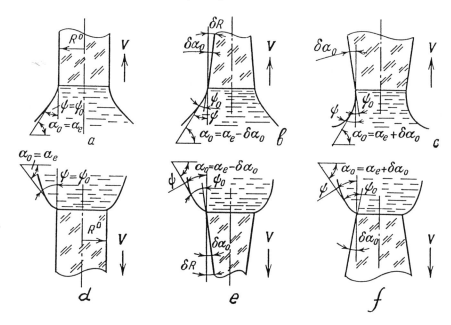

Fig. 1.8.   Crystal growth for the case of capillary shaping: $\alpha_0 = \alpha_e$-growth of crystal of constant cross-sections (a, d); $\alpha_0 < \alpha_e$-growth of crystal of widening cross-sections (b, e); $\alpha_0 > \alpha_e$-growth of crystals of narrowing cross-sections (c, f); V is crystal displacement rate.

The angle α_0 as a function of crystal dimensions, crystallization front position and other parameters together with the meniscus shape can be determined by solving the capillary problem whose equation was discussed in 1.2. Assuming that the capillary problem is solved, i.e., the function $\alpha_0(X_1, X_2, \ldots, X_k, \ldots \ldots, X_n)$ is found, (1.38) can be represented in the following form:

$$\delta \dot{R} = -V \sum_{k=1}^{n} (\partial \alpha_0 / \partial X_k) \delta X_k \qquad (1.39)$$

Here $\delta X_k = X_k - X_k^0$, X_k^0 is the value of the parameter X_k with $\alpha_0 = \alpha_e$. Thus the first equation (for the crystal cross-section change rate) of (1.3) is obtained.

In the following chapters much consideration will be given to the formulation and solution of the capillary problems for the capillary-shaping techniques discussed here.

1.6. Temperature Distribution in the Crystal-Melt System

The second equation of (1.3), the equation of the crystallization-front displacement rate, is obtained by solving the thermal conductivity problem. The problem of defining temperature distributions in the growing crystal and in the melt in the area adjacent to the crystallization front will be called the thermal conductivity problem. There exists a great number of works dealing with calculation of the temperature fields in the crystal-melt system. They form a group of the Stefan problems [87], the problems with a moving boundary which is a heat source. However, owing to the variety of growth schemes and a number of factors that need to be taken into account in the thermal conductivity problems, such as complex temperature dependence of the thermophysical characteristics of various matters, a complete mathematical description of heat patterns during crystal growth is not yet available. Obtaining the solution in its analytical form is usually achieved by significant simplifications, i.e. by assigning some specific temperature dependence of the transfer coefficients. Thus, the thermal conductivity coefficient is assumed to be proportional to T^{-1} in [88] and to T^3 in [89], in [33] linearization of the law of crystal-surface radiation is proposed. Recently, numerical methods have often been used to solve such problems.

Investigation of the relations between the technological parameters of the crystallization process, growing-crystal dimensions and the crystallization front position requires simultaneous consideration of the heat flows in the crystal-melt system, i.e., simultaneous solution of the convective heat equation and the Navier-Stokes equation allowing for the flows associated with both melt crystallization and convection. However, while analyzing crystallization process stability, the signs of the derivatives of the temperature gradients at the crystallization front for the respective parameters (crystal radius, crystallization front position, etc.) will be sufficient. With this end in view the following equation allowing simple analytical solutions [87] will be used to analyze the heat conditions in the crystallization process:

$$\frac{1}{k_i}\frac{\partial T_i}{\partial t} = \frac{d^2 T_i}{dz^2} - \frac{V}{k_i}\frac{dT_i}{dz} - \frac{\mu_i}{\lambda_i}F(T_i - T_e) \tag{1.40}$$

Here $i = L, S$ ($i = L$ for a liquid, $i = S$ for a solid body), T_i denotes the temperature, k_i is the thermal diffusivity coefficient, z is the vertical coordinate, μ_i denotes the coefficient of heat-exchange with the environment, F denotes the crystal (meniscus) cross-section perimeter-to-its area ratio, T_e is the environment temperature, λ_i is the thermal conductivity coefficient.

Then (1.40) describes the temperature distribution of the crystal-melt system in a one-dimensional approximation. This means that the temperature in the crystal (meniscus) cross-section is averaged and real isotherms are replaced with flat ones. In this case, the heat exchange with the environment is allowed for not in the form of the boundary condition as in the case of multi-dimensional problems but by introducing heat run-offs on the lateral surface in the form of an additional term in the heat equation.

Then (1.40) gives a good description of real temperature distribution for small Biot numbers ($Bi = \mu_i R/\lambda_i \ll 1$). This can be observed during growth of small diameter crystals (characteristic of CST)[1], for low coefficients of convective heat transfer from

---

[1]For this reason, the Kyropoulos technique used to grow cyrstals of a large diameter will not be considered here. In fact, it is also classified as a capillary-shaping technique.

the crystal (melt) surface, μ_i, and high thermal conductivities, λ_i, (characteristic of metals). However, if Biot numbers are not too small (1.40) gives a qualitatively correct description of the dependence of the temperature gradients in the melt, G_L, and in the crystal, G_S, on the crystal dimensions, crystallization front position, pulling rate and some other parameters.

The interface temperature is equal to the crystallization temperature, T_0. In some cases T_0 dependence on the crystallization rate, V_C, should be taken into consideration. This effect will be analyzed below. Other boundary conditions depend upon specific crystallization schemes and cooling conditions.

While growing single crystals of transparent substances, radiance flow availability is most often equalized by introducing the effective thermal-conductivity coefficient proportional to T_i^3 [89]. In this case, if heat exchange on the lateral surface occurs by radiation following the Stefan-Boltzmann law, then an equation of the (1.40) type for T_i^4 is obtained. Therefore, the results obtained for (1.40) can be used in this case too with T_i replaced by T_i^4.

Heat exchange is allowed for by the Newton law in case the convective heat exchange is much higher than the heat losses caused by radiation. As shown in [90], the heat losses caused by free convection are comparable with the heat losses caused by radiation at the surface temperatures of $\sim 1000^0 C$ and even higher where specimen surface blowing is provided. Then (1.40) can be used for low-melting materials, whereas in the case of refractory materials, especially when vacuum pulled, radiation heat-exchange (the Stefan-Boltzmann law) needs to be allowed for, which leads to considerable non-linearity of the problem. As shown in [33], linearization of the crystal-surface radiation law leads to reasonable results that agree well with the ruby growing experiment, i.e., for the temperatures of 2100^0C. Linearization is performed as follows. Heat losses from a unit crystal surface per unit period of time Q can be represented in the following way:

$$Q = \mu_k(T_i - T_e) + \sigma(T_i^4 - T_e^4)f_{cw}$$

Here μ_k is the convective heat-exchange coefficient, σ is the Stefan-Boltzmann constant, T_e is both the environment temperature and the temperature of the screens absorbing the radiation, f_{cw} is the factor allowing for the blackness and the shape of the crystal and the walls absorbing the radiation. Now we will rearrange the previous equation in the following way:

$$Q = [\mu_k + \sigma f_{cw}(T_i^4 - T_e^4)/(T_i - T_e)](T_i - T_e)$$

By replacing the variable temperature T_i in the square brackets by the mean radiating-surface temperature T_S, we obtain:

$$\mu = \mu_k + \mu_r \tag{1.41}$$

where μ_r is the radiation heat-exchange coefficient:

$$\mu_r = \sigma f_{cw}(T_S^2 + T_e^2)/(T_S + T_e) \tag{1.42}$$

In the case of ruby growing by the Verneuil technique it is assumed in [33] that $\mu = 500 J m^{-2} s^{-1} K^{-1}$.

Thus, the problem formulation offered is general and in particular cases one of the coefficients can be neglected. By introducing a new variable $T_i^* = T_i - T_e$, Equation (1.40) can be simplified:

$$\frac{1}{k_i} \frac{\partial T_i^*}{\partial t} = \frac{dT_i^*}{dz^2} - \frac{V}{k_i} \frac{dT_i^*}{dz} - \frac{\mu_i}{\lambda_i} F T_i^* \tag{1.43}$$

1.7. The Equation of the Crystallization Front Displacement Rate

Now we will derive another equation of (1.3) that is valid for all the capillary-shaping techniques, i.e., the equation of the crystallization front displacement rate [1, 91, 51]. At the crystallization front the heat-balance equation should be satisfied:

$$-\lambda_S G_S(h) + \lambda_L G_L(h) = \mathcal{L} V_C \tag{1.44}$$

Here λ_S and λ_L denote thermal conductivities of the solid and liquid phases respectively, $G_S(h)$ and $G_L(h)$ are the temperature gradients of the solid and liquid phases at the crystallization front, h is a crystallization front coordinate; $\mathcal{L}$ denotes the latent melting heat of a material unit volume, V_C is the crystallization rate. In this case, if we use V to denote the pulling rate then in accordance with (1.37) $V_C = V - dh/dt$. Thus, from (1.44) we obtain an (1.1)-type equation for dh/dt:

$$dh/dt = V - \mathcal{L}^{-1}(\lambda_L G_L - \lambda_S G_S) \tag{1.45}$$

Now the expressions for G_L and G_S are to be found, which can be done using (1.40) with corresponding boundary conditions. However, this rigid approach presents great mathematical difficulties. As was mentioned above, these difficulties can be overcome using a quasi stationary approximation according to which temperature distribution in the crystal-melt system at any moment of time satisfies the stationary thermal conductivity equation with instantaneous values of all the process parameters. For this approximation to be applied, the temperature in the system should reach the stationary value much faster than the crystallization front relaxes. In other words, the time of crystallization front relaxation, τ_{rf}, to the stationary state should be significantly longer than the characteristic time of temperature relaxation, τ_{rT}, to the state corresponding to the specified values of the process parameters

$$\tau_{rf} \gg \tau_{rT} \tag{1.46}$$

Then, as in [87], for the system for which temperature distribution is described by (1.40)

$$\tau_{rT}^{-1} = k_i \pi / L^2 + V/2k_i^2 + 2\mu_i k_i F/\lambda_i \tag{1.47}$$

Here L denotes the characteristic system dimension, while other designations are the same as in (1.40). Numerical calculations illustrating the validity of (1.46) for most of the cases will be given below.

Assume that the solution of the stationary problem (the boundary problem for (1.40) with the left side equal to zero for corresponding boundary conditions) results in calculation of liquid- and solid-phase temperature distribution. Then the temperature is a function of the boundary conditions, the crystallization front position, h^0, being one of them. Liquid- and solid-phase temperature gradients will be functions of the same boundary conditions and the heat balance equation (1.44) should be satisfied at the crystallization front; then $V = V_C$ as the crystallization front displacement rate is equal to zero due to the stationary process.

Assume that some of the parameters X_k of our system change by a small value $\delta X_k = X_k - X_k^0$. If Equation (1.46) is valid for the system, new temperature gradients in (1.44) will be found from the stationary problem as functions of the new parameters X_k, and the crystallization front will move with the rate $\delta \dot{h} = d(h - h^0)/dt$ towards a new equilibrium position. By expanding the temperature gradients into Taylor series in terms of δX_k in the vicinity of the stationary point and using linear terms δX_k of expansion and taking into consideration that the temperature gradients with $X_k = X_k^0$ are the solutions of (1.44) with $\delta \dot{h} = 0$ (the non-perturbed problem), we obtain:

$$\delta \dot{h} = \mathcal{L}^{-1} \sum_{k=1}^{n} [\lambda_S(\partial G_S/\partial X_k) - \lambda_L(\partial G_L/\partial X_k)]\delta X_k \tag{1.48}$$

Thus, we have the second equation of (1.3) for the crystallization front displacement rate. It is more convenient to consider the explicit form of other equations of (1.3) for each crystallization technique separately.

1.8. Convection Influence on the Liquid-Phase Heat-Transfer Processes

As a rule, the crystallization process is carried out under conditions of liquid-phase temperature gradients. As is known, availability of such a gradient leads to convective flow formation, to free convection in particular, associated with gravity presence that is characterized by the Raleigh number, $Re = \Delta T g \beta L^3 (\eta k)^{-1}$, and by the so-called Marangoni convection associated with the surface-tension gradient characterized by the Marangoni heat number, $Ma = \frac{(d\gamma_{LG})}{dT} \frac{(\Delta T L)}{\eta k}$. Here β denotes the volume expansion coefficient, ΔT denotes the characteristic temperature difference, η is the coefficient of dynamic viscosity.

The number $(g\beta L)^2 (d\gamma_{LG}/dT)^{-1}$, that is the Raleigh number-to-Marangoni number ratio, characterizes relative and capillary-force effects on convective flow formation. For small characteristic dimensions of the system the capillary forces prevail. In [92] it is experimentally shown that the Marangoni convection develops in the zone with the characteristic dimensions $L \approx 5$ mm. As the characteristic dimensions of the meniscus in the capillary-shaping techniques are of the order of a few millimeters, the influence of this very convection on the heat-transfer processes is discussed here [61].

If the melt is overheated, the surface tension at the crystallization front is higher and hence the melt moves towards the front on the meniscus-free surface and off the crystallization front in the center under the action of the pressure gradient formed. When crystals grow in the upward direction, the pressure gradient mentioned is converse to the pressure gradient associated with the gravity field present. For supercooled-melt growing the opposite can be observed.

Liquid-phase convection results in melt stirring and in heat removal increase at the crystallization front for overheated-melt growing. The magnitude of the rates of this flow can be estimated from the meniscus free-surface boundary condition that can have the following form

$$\eta dV/dr = -(d\gamma_{LG}/dT) \text{ grad } T \tag{1.49}$$

Hence,

$$V \approx \frac{d\gamma_{LG}}{dT} \frac{R\Delta T}{\eta h} \tag{1.50}$$

where R denotes the crystal radius (meniscus cross-section), h denotes the meniscus height. The Marangoni heat number is characterized by the ratio of the melt motion-transferred heat flow with the rate V to the diffusion-transferred heat flow. Rate estimation from the above expression gives the values an order of magnitude higher than those observed in [92]. This difference is explained by the frictional effect of the crystal solid surface. As is noted in [92], convection contribution to the heat transfer is considerable for liquid zones more than 1mm long. Note that the convective heat flow towards the crystallization front increases with R increasing and h decreasing, which can be seen when estimating the magnitudes of the convective flow rates. The behavior of the diffusion-transferred heat flow is the same.

Thus, within crystallization process stability study, convection influence on the heat transfer can be allowed for by formal introduction of the effective coefficient of thermal conductivity and that of lateral-surface heat exchange.

Chapter 2

THE CZOCHRALSKI TECHNIQUE

2.1. The Lyapunov Set of Equations

If the crystal dimension (its radius) R and the crystallization front position relative to the melt-free surface h are regarded as variable parameters in the Czochralski technique, for stability study (1.3) will include (1.39) and (1.48):

$$\delta \dot{R} = A_{RR} \delta R + A_{Rh} \delta h \tag{2.1}$$

$$\delta \dot{h} = A_{hR} \delta R + A_{hh} \delta h \tag{2.2}$$

The following notations are used here:

$$A_{RR} = -V(\partial \alpha_0 / \partial R) \tag{2.3}$$

$$A_{Rh} = -V(\partial \alpha_0 / \partial h) \tag{2.4}$$

$$A_{hR} = \mathcal{L}^{-1}[\lambda_S(\partial G_S / \partial R) - \lambda_L(\partial G_L / \partial R)] \tag{2.5}$$

$$A_{hh} = \mathcal{L}^{-1}[\lambda_S(\partial G_S / \partial h) - \lambda_L(\partial G_L / \partial h)] \tag{2.6}$$

G_S and G_L denote the solid- and liquid-phase temperature gradients, respectively. The solutions of the system (2.1), (2.2) are:

$$\delta R = C_1 \exp(S_1 t) + C_2 \exp(S_2 t) \tag{2.7}$$

$$\delta h = C_3 \exp(S_1 t) + C_4 \exp(S_2 t) \tag{2.8}$$

where S_1 and S_2 are roots of the characteristic equation

$$S^2 - (A_{RR} + A_{hh})S + (A_{RR}A_{hh} - A_{Rh}A_{hR}) = 0 \tag{2.9}$$

45

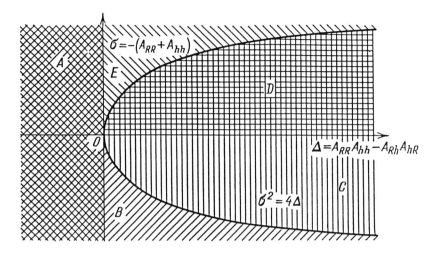

Fig. 2.1.   Areas and types of stability: A – saddles, B – unstable nodes, C – unstable focuses, D – stable focuses, E – stable nodes.

For the system stability it is necessary and sufficient that the real parts of the characteristic equation roots S_1 and S_2 should be negative. Then perturbations will attenuate with time. However, to estimate the stability of the set of equations it is not necessary to solve the equations, instead the Routh-Gurvitz conditions [60] can be used. As follows from these conditions, for the set of two equations (2.1) and (2.2) to be stable it is necessary and sufficient that the equation coefficients should satisfy the following inequalities:

$$A_{RR} + A_{hh} < 0 \tag{2.10}$$

$$A_{RR}A_{hh} - A_{Rh}A_{hR} > 0 \tag{2.11}$$

With the conditions (2.10) and (2.11) satisfied, two types of stable behavior of the system are possible depending upon the sign of the inequality

$$4A_{Rh}A_{hR} \lessgtr (A_{RR} - A_{hh})^2$$

The sign $<$ corresponds to the point called "a node" on the phase diagram, while the sign $>$ corresponds to the point called "a focus" (Figure 2.1). The focus corresponds to the system's absolute stability.

The coefficients A_{RR} and A_{hh} indicate direct correlation between $\delta \dot{R}$ and δR, $\delta \dot{h}$ and δh, i.e., self-stability of the parameters. The coefficients A_{Rh} and A_{hR} represent the effect of one value change on the other value change rate, i.e., interstability of the parameters. To achieve interstability in the system, the inequality $A_{Rh}A_{hR} < 0$ should be satisfied, i.e., the coefficients A_{Rh} and A_{hR} should have different signs (see (2.11)). If $A_{RR} > 0$ and $A_{hh} > 0$ the system will be unstable as follows from

(2.10) since self-stabilization of both the parameters is absent. In case $A_{RR}A_{hh} < 0$, i.e., self-stabilization of one of the parameters is absent, system stability is possible if interstabilization is available.

The stability types mentioned are "rough" in the sense that the system stability remains unchanged within a wide range of values of the coefficients $A_{jk}(j, k = R, h)$. If at least one of the inequalities (2.10), (2.11) is replaced by an equality, the roots of the characteristic equation (2.9) are either imaginary or comprise zero ones. In this case, we cannot judge the system stability by its linear approximation, rather a nonlinear model should be analyzed.

To define the explicit form of the coefficients A_{jk} specified by (2.3)–(2.6) the capillary problem and the heat problem for crystal growth by the Czochralski technique should be solved.

2.2. Liquid Meniscus Height and Shape (The Capillary Problem)

The first attempt to define the growing crystal diameter as a function of the melt meniscus height was made in [93]. This attempt was not a success and the results are given here to warn against such mistakes. Here, the true liquid meniscus is substituted by a straight cylinder of the radius R and the height h representing an extension of the growing crystal. The weight of the cylinder is equated to the force of surface tension at the liquid-solid phase boundary:

$$2\pi R \gamma_{LG} = \pi R^2 h \rho_L g$$

hence

$$h = (2\gamma_{LG}/\rho_L g)R^{-1} = a^2 R^{-1}$$

The crystallization front height h turns out to be inversely proportional to the growing crystal radius R. This relationship is not proved by experimental results. The author's mistake is in equating the force of surface tension to the weight of the cylindrical liquid meniscus, in fact it places only the weight of the liquid outside the cylindrical part of this meniscus in equilibrium. By equating these forces the Laplace capillary equation describing the liquid meniscus shape can be obtained.

In all other works discussed below the Laplace capillary equation is used to define the liquid meniscus shape. In its generalized form this equation cannot be solved. Therefore all the works are mainly dedicated to various methods of its approximated solution.

2.2.1. Problem Equation

If the z-coordinate origin lies on the melt-free surface, $const = 0$ in (1.9) and then (1.12) has the following form:

$$z''r + z'(1 + z'^2) - 2z(1 + z'^2)^{3/2}r = 0 \tag{2.12}$$

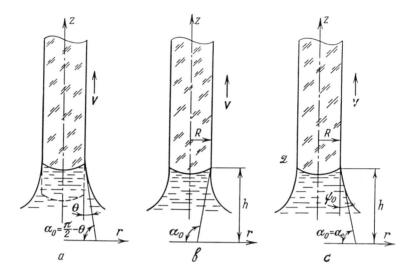

Fig. 2.2.   Subsequent seed position when raised out of the melt (a, b) and growth of crystals of constant cross-sections (c); L-melt-crystal conjugation contour.

2.2.2. *Boundary Conditions*

The lower-boundary condition is determined by crystal growth from the melt-free surface:

$$z\,|_{r\to\infty} = 0 \tag{2.13}$$

The crystal-melt boundary condition. If a seed is sunk into the melt, the melt will rise along it and make an angle equal to the wetting angle θ with it (Figure 2.2a), i.e., the wetting condition is satisfied at the meniscus upper edge at the starting moment.

The term "the wetting boundary condition" was introduced by us [29–31] and is now widely used in the literature in a broader sense than was originally intended. Now "the angle fixation condition" for the angle made by the line tangent to the melt surface with the direction specified without an allowance for the angle fixation technique is referred to as the wetting condition. However, this can lead to some misunderstanding. For example, according to this terminology the wetting boundary condition at the crystal-free edge in the process of growing crystals of constant cross-sections is specified; in this case, the line tangent to the melt is supposed to make an angle equal to the growth angle with the vertical at the crystal-free edge. In such a way, the wetting angle and the growth angle can be confused. Hence, we will use the term "the angle fixation condition" for (2.14). Only where the angle is fixed on the shaper wall will the term "the wetting condition" be used.

If the seed is pulled upwards, the melt slides downwards and catches on the seed lower free edge, i.e., the catching condition is satisfied (Figure 2.2b).   The term "catching" was also introduced by us [29–31]. Any case where the position of some line on the meniscus surface is fixed (along the contour $\hat{\mathcal{L}}$ in Figure 2.2b) will be referred to as "catching" from now on.

However, to grow a vertical-walled crystal, the melt should make a fixed angle with the growing crystal surface equal to the growth angle ψ_0 at the point of contact with the growing crystal during crystallization (Figure 2.2b). Hence, while growing a crystal of constant cross-sections the angle-fixation condition should be satisfied at the melt-crystal contact point (Figure 2.2c).

$$z' \mid_{r=R} = -\text{tg}\alpha_0,\, \alpha_0 = \pi/2 - \psi_0 = \alpha_e \tag{2.14}$$

The case of anisotropic growth when shaping occurs under the melt surface and the melt makes the wetting angle with a tightly-packed facet of the crystal lateral surface is not discussed here.

It should be emphasized again that for all the capillary shaping techniques the condition (2.14) is only satisfied during constant cross-section crystal growth while the catching condition is always satisfied at the crystal-free edge (Figure 2.2b, c):

$$z \mid_{r=R} = h \tag{2.15}$$

2.2.3. Analytical-Numerical Solution

Any numerical solution technique is difficult to apply to (2.12) because of (2.13). However, a combined solution can be obtained: a numerical one for the r-range from R to rather large r_m values and then an asymptotical part of the solution (analytical one) from r_m to $r\infty$ [94].

In our case $z' < 0, z'^2 \ll 1$ for $r > r_m$. Substitution of $\zeta = \sqrt{2}z, y = \sqrt{2}r$ gives the following form of (2.12) (the Nicolson approximation [95]):

$$\zeta'' - y^{-1}\zeta' - \zeta = 0 \tag{2.16}$$

This is a differential equation, its solution being the modified Bessel's function of the second kind:

$$\zeta(y) = A/K_0(y) \tag{2.17}$$

The solution of (2.17) satisfies the boundary condition (2.13). The constant A can be defined from the condition of the first-derivative continuity at the numerical-analytical solution conjugation point with the coordinates y_m and ζ_m:

$$A = [y'(\zeta_m)K_1(y_m)]^{-1} \tag{2.18}$$

Here K_1 also denotes the modified Bessel's function of the second kind.

For numerical solution of the problem in the range of $[R, r_m]$ the boundary conditions (2.14), (2.15) can be used. In [94] the Runge-Kutta technique was employed and

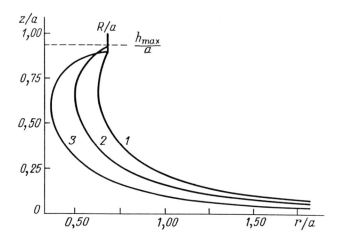

Fig. 2.3.   Profile curves (meniscus section by the diagram plane) for $R/r = 0.7$: 1 – stable meniscus, $h/a = 0.882$, $\alpha_0 = 144.6°$; 2 – meniscus with the maximum height, $h_{max}/a = 0.933$, $\alpha_{0\,max} = 151.57°$; 3 – unstable meniscus, $h/a = 0.882$, $\alpha_0 = 172°$ (a is the capillary constant) [94].

numerical-analytical solution conjugation was provided at the point of $\zeta' \leq 0.005$, which led to a systematic error for ζ'^2 not exceeding $2.5 \cdot 10^{-5}$ when using the analytical technique.

If both positive and negative ranges of the angles ψ_0 are considered, α_0 will vary from 0 to 180°. In this case, h will increase with α_0 increasing, it will reach its maximum value, h_{max}, at some $\alpha_{0\,max}$ and then it will decrease (Figure 2.3). It therefore follows that two profile curves with differing values of $(\alpha_{01} > \alpha_{02})$ will correspond to the fixed point (R, h) with $h < h_{max}$ (Figure 2.3, curves 1, 3).

2.2.4. Static Stability of Meniscus

As was noted in 1.2, for stable meniscus the Jacobi equation should not have conjugate points and the Legendre condition should be satisfied.

Since axially symmetric melt columns are formed during crystal growth by the Czochralski technique, we will use the cylindrical coordinate system $r^*\varphi w$ for which the functional (1.6) will have the following form:

$$C \int (\gamma_{LG} 2r^* (1 + r^{*'2})^{1/2} + \rho g w r^{*2}) dw = \min$$

while in the dimensionless coordinates $z = w/a, r = r*/a$, it will assume the following form:

$$C \int (r(1 + r'^2)^{1/2} + r^2 z) dz = \min \tag{2.19}$$

Thus, the function being minimized in our problem has the form:

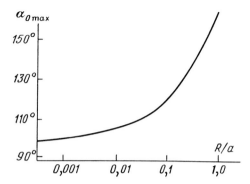

Fig. 2.4. Maximum value of the angle $\alpha_{0\,max}$ corresponding to the boundary of statically stable meniscus as a function of crystal radius [94].

$$f(z, r, r') = C[r(1 + r'^2)^{1/2} + r^2 z] \tag{2.20}$$

And since $\partial^2 f/(\partial r')^2 = Cr(1 + r'^2)^{-3/2} > 0$, according to the Legendre condition we obtain the minimum of (2.20). The explicit form of the Jacobi equation for the Czochralski technique is obtained from the general expression (1.16):

$$\eta'' - \frac{\sqrt{2}r'}{r}\left[2 + \frac{3}{2}rz(1 + r'^2)^{1/2}\right] + \frac{2}{r^2}(1 + r'^2)\eta = 0 \tag{2.21}$$

r'' is excluded from (2.21) by using the capillary equation (2.12). In [94] (2.21) was numerically solved. It was found that conjugate points appear for $\alpha_0 > \alpha_{0\,max}(\alpha_{0\,max} > 90°)$ corresponding to the maximum height of the meniscus (Figure 2.3). Whence it follows that the relation $\alpha_{0\,max}(R)$ should state whether or not the meniscus will be stable. Such a relation is plotted in Figure 2.4. Note that since the whole range of $\alpha_{0\,max}$-values are obtuse angles, the growth angles $\psi_0 < 0$ corresponding to them are negative, i.e., negative growth angles meniscus should be analyzed for static stability. Figure 2.5a, b shows stable meniscus for $R/a = 0.7$.

2.2.5. Approximated Analytical Solution

Some analytical approximated solutions of the Laplace capillary equation will now be considered. If this equation for axially symmetric melt columns is written down in the form of (1.5), one of the main curvature radii, R_1, will lie within the diagram plane (Figure 2.6), the other one, R_2, will lie within the plane perpendicular to it. In this case:

$$1/R_2 = [r(1 + r'^2)^{1/2}]^{-1} \tag{2.22}$$

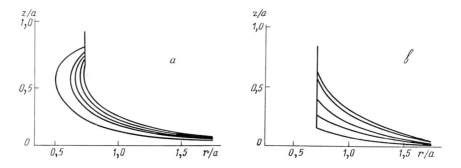

Fig. 2.5. Profile curves for a crystal of the radius $R = 0.7$ for various growth angles: (a) $\psi_0 \leq 0$; (b) $\psi_0 \geq 0$.

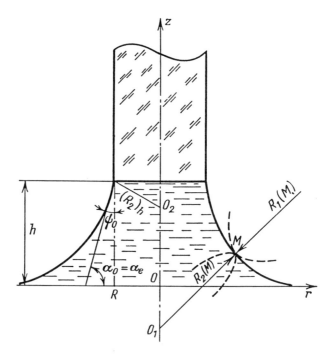

Fig. 2.6. Main meniscus parameters for growing crystals by CST: $R_1(M), R_2(M)$ – principal curvature radii at an arbitrary point M; $(R_2)_h$ – the principal curvature radius at the crystal-free edge; ψ_0 – growth angle; α_e – complement of the growth angle; α_0 – the angle made by the line tangent to the melt on the crystal free surface with the horizontal.

In [71] $1/R_2$ is represented in the form of the Taylor series:

$$\frac{1}{R_2} \left(\frac{1}{R_2} \right)_h + \left(\frac{d}{dz} \frac{1}{R_2} \right) (z - h) + \ldots \tag{2.23}$$

Then, it is proposed to regard $1/R_2$ as a linear function of the vertical coordinate z:

$$\left(\frac{d}{dz}\frac{1}{R_2}\right)_h = \frac{1}{h}\left[\left(\frac{1}{R_2}\right)_h + \left(\frac{1}{R_2}\right)_0\right] \tag{2.24}$$

As follows from simple geometrical relationships (Figure 2.6),

$$\left(\frac{1}{R_2}\right)_h = \frac{\sin\alpha_0}{R}, \qquad \left(\frac{1}{R_2}\right)_0 = 0 \tag{2.25}$$

Substitution of these expressions into (2.23) allows its reduction to the following form:

$$\frac{1}{R_2} = \frac{\sin\alpha_0}{Rh}z \tag{2.26}$$

With (2.26) substituted into (1.5) and the relation $1/R_1 = r''(1 + r'^2)^{-3/2}$ taken into consideration, (1.5) can be integrated. As a result, in [71] an expression for the melt column height during crystal pulling from the melt-free surface was obtained:

$$h(R,\alpha_0) = \sqrt{1 - \cos\alpha_0 + \frac{\sin^2\alpha}{16R^2}0 - \frac{\sin\alpha}{4R}0} \tag{2.27}$$

A particular case of (2.27) for $\alpha_0 = \pi/2$ is given in [67]. Here the following relationship is obtained for a growing crystal of a large diameter:

$$h(r,\pi/2) = 1 - 1/4R, \qquad R \geq 1/4 \tag{2.28}$$

For small growth angles $\psi_0 \to 0$, $\alpha_0 \to \pi/2$ (2.27) can be simplified:

$$h(R,\psi_0) = h(R,\pi/2) - (1 + 1/16R^2)^{-1/2}\psi_0 \tag{2.29}$$

Figure 2.7b represents families of curves $h(\psi_0)$ for various R-values plotted analytically using (2.27) and as a result of numerical calculations (Figure 2.7a) [96]. It is obvious that for $\psi_0 \geq 0$ and large R-values this formula is applicable, while for $\psi_0 \to -90°$ it is not valid. It can be attributed to the progressing error appearing in the approximation of (2.24) with α_0 increasing.

For $R \ll 1$ (the diameter of the crystal being pulled is less than the capillary constant) (2.27) can be rearranged in the follow way:

$$h(R,\alpha_0) = \frac{2R(1 - \cos\alpha_0)}{\sin\alpha_0}, \qquad R \ll 1 \tag{2.30}$$

With $\psi_0 = 0(\alpha_0 = \pi/2)$, the melt column height turns out to be equal to the diameter of the crystal being grown.

For the case of pulling a crystal of a small diameter, approximation [97] can be used:

$$h(\alpha_0,R) = R\sin\alpha_0\left[\ln\frac{4}{R(1 - \cos\alpha_0)} - \vartheta\right], \qquad R < 1 \tag{2.31}$$

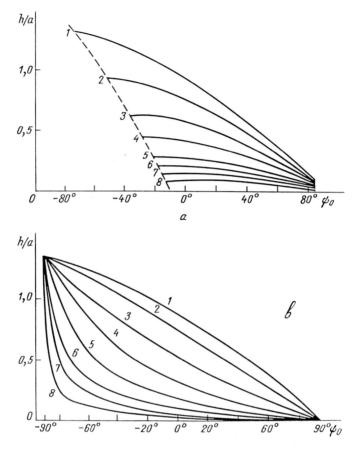

Fig. 2.7.   Meniscus height versus growth angle for various crystal radii R/a: $1 - 1.414$; $2 - 0.707$; $3 - 0.354$; $4 - 0.177$; $5 - 0.088$; $6 - 0.044$; $7 - 0.022$; $8 - 0.011$. The dash line limits the area of stable meniscus. Numerical calculation (a), calculated from (2.27) (b).

Here $\vartheta = 0.57712\ldots$ is the Euler constant. This formula is also given in [98].

Approximations [99, 100] are applicable to growing crystals of large diameters. The Laplace capillary equation for the function $r(z)$ is used in the form of (1.15):

$$\frac{r''}{(r'^2 + 1)^{3/2}} - \frac{1}{r(r'^2 + 1)^{1/2}} - 2z = 0 \tag{2.32}$$

The second term of (2.32) is considered to be negligible. Integration of the equation obtained is used as the first approximation. Using this approximation the second term neglected at the beginning can be calculated. As a result, the following equation for $h(\pi/2, R)$ is given in [99]:

$$h^2 - 1 + \frac{4}{3\sqrt{2}R}\left[1 - \left(1 - \frac{h^2}{2}\right)^{3/2}\right] = 0 \tag{2.33}$$

Further simplification is used in [100]:

$$(1 - h^2/2)^{1/2} \approx 1 \tag{2.34}$$

hence,

$$h(\pi/2, R) = (1 + 1/\sqrt{2}R)^{-1/2} \approx 1 - 0.35/R \tag{2.35}$$

A very useful definition of the applicability ranges for all the approximations described by comparing them with the results of numerical calculations made according to the procedure mentioned above is cited in [96][1].

Figure 2.8 shows the relationship $h(\psi_0)\ |_{R=const}$. It can be seen that approximations from [100] and [97] both give the maximum height for $\psi_{0\,max} > -90°$ in contrast to approximation from [71]). With $R \sim 1$ good agreement is observed for approximation from [100] while for approximation from [97] it is absent as was to be expected. The opposite is observed for small R-values (Figure 2.8b, $R \sim 0.01$) when approximation from [97] fits very well.

Figure 2.9 shows the relationship $h(R)\ |_{\alpha_0=const}$ for the accurate numerical solution and for all the approximations. Figure 2.10 represents deviations from the accurate numerical solution when applying each of the approximations. The range of approximation applicability can be easily defined. For $R > 1$ approximation from [99] is the best to use.

The asymptotic formula for approximations from [100, 99, 67] can be written in the following form:

$$h\left(R, \frac{\pi}{2}\right) = 1 - K\frac{1}{R} + O\left(\frac{1}{R^2}\right) \tag{2.36}$$

where $K = 0.251$ for approximation from [67], $K = 0.306$ for [99], and $K = 0.354$ for [100].

In [101] for the first integral of the Laplace capillary equation it is suggested that the expression including the modified Bessel's functions should be used

$$z = (1 + \cos\alpha)^{1/2} K_0(\sqrt{2}r)/K_1(\sqrt{2}r) \tag{2.37}$$

In this case the ratio of the Bessel's functions is replaced with the following expression:

$$K_0(r)/K_1(r) \approx (1 + r^{-1})^{-1/2} \tag{2.38}$$

Equation (2.38) gives an error of 3.3% for $r = 0.5$; 0.27% for $r = 2$ and 0.01% for $r = 10$.

Our review would not be exhaustive if we did not mention the monograph [102], where everything that was known about capillarity in the thirties is collected. Besides approximation from [99] considered above, the following relationships from this monograph can be useful for us:

---

[1]Uelhoff and Mika [96] use the approximation from Games [98] although in fact this result was published eleven years earlier in Voronkov [97].

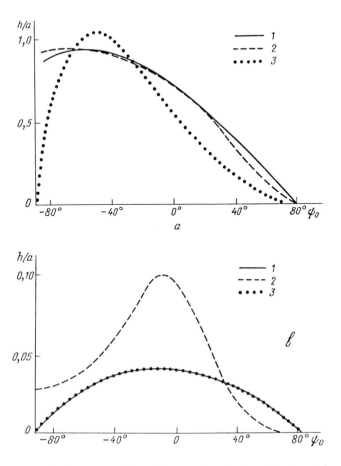

Fig. 2.8.   Meniscus height vs. growth angle for crystals of various diameters, $R/a = 0.7$ (a), $R/a = 0.0007$ (b): 1 – numerically calculated, 2 – Heywang's approximation, 3 – Voronkov's approximation [96].

(a)   The liquid rise height h from the free surface along a wettable plate as a function of the wetting angle θ:

$$h = a\sqrt{1 - \cos \theta} \tag{2.39}$$

(b)   With the plate wettability being absolute ($\theta = 0$), the liquid rise height reaches its maximum and is equal to the capillary constant a.

(c)   The equation of the profile curve $z(r)$ for liquid rising from its free surface along a wettable plate is:

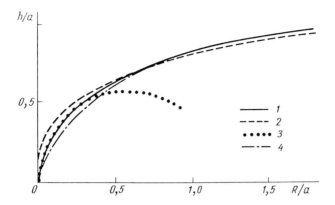

Fig. 2.9. Meniscus height vs. crystal radius for zero growth angle: 1 – numerical calculation, 2 – Heywang's approximation, 3 – Voronkov's and Games' approximations, 4 – Gaule-Pastore's and Tzivinskiy's approximations [96].

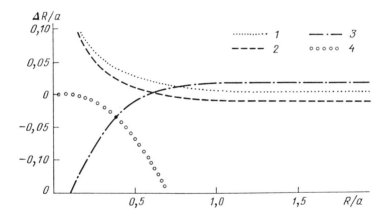

Fig. 2.10. Deviation of meniscus-height approximated values from the accurate ones as a function of crystal radius: 1 – Ferguson's approximation, 2 – Heywang's one, 3 – Gaule-Pastore's and Tzivinskiy's ones, 4 – Voronkov's and Games' ones [96].

$$r + C^* = (2a^2 - z^2)^{1/2} - \frac{a}{\sqrt{2}} \ln \frac{z}{a\sqrt{2} - (2a^2 - z^2)^{1/2}} \qquad (2.40)$$

C^* is a constant determined by the wetting angle and by the plate slope towards the liquid surface. A particular case of this formula (for a vertical plate) is represented by (1.22).

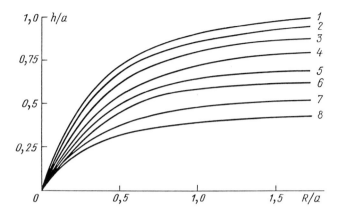

Fig. 2.11.   Profile curve height vs. crystal radius for various growth angles ψ_0: (1) $-20°$; (2) $-10°$; (3) $0°$; (4) $10°$; (5) $20°$; (6) $30°$; (7) $40°$; (8) $50°$.

The monograph [103] and the paper [104] considering a number of interesting capillarity problems and tables [105, 106] for numerical calculation of examples of the Laplace capillary equation should also be mentioned. In [107] detailed instructions on applying the tables are given.

In [96] mentioned above, detailed tables for plotting profile curves of various-diameter crystal growth by the Czochralski technique are given.

The relationships $h(R)|_{\alpha=const}$ (Figure 2.11), $\frac{\partial \alpha_0}{\partial h}(R)$ and $\frac{\partial \alpha_0}{\partial h}(R)$ (Figure 2.12) were plotted using these tables.

2.2.6. Analysis of Capillary Coefficients

While studying stability, we will need the relationships $\frac{\partial \alpha_0}{\partial h}(R)$ and $\frac{\partial \alpha_0}{\partial h}(R)$ (Figure 2.12). The coefficients $A_{RR}(R)$ and $A_{Rh}(R)$ are obtained from $\frac{\partial \alpha_0}{\partial R}(R)$ and $\frac{\partial \alpha_0}{\partial h}(R)$ by the formulae (2.3) and (2.4). From Figure 2.12 it follows that the values of $\frac{\partial \alpha_0}{\partial R}$ remain negative for all R-values (i.e., $A_{RR} > 0$), approaching zero for crystals of large diameters. The values of $-\frac{\partial}{\partial h}\frac{\alpha}{h}\bar{0}(R)$ are positive for all R-values (i.e., $A_{Rh} < 0$). In order that the stability conditions (2.10), (2.11) could be satisfied with the sign of the coefficients A_{RR} and A_{Rh} obtained, the following inequalities should be satisfied:

$$A_{hh} < 0, \qquad |A_{hh}| > |A_{RR}| \tag{2.41a}$$

$$A_{hR} > 0, \qquad |A_{Rh}||A_{hR}| > |A_{RR}||A_{hh}| \tag{2.41b}$$

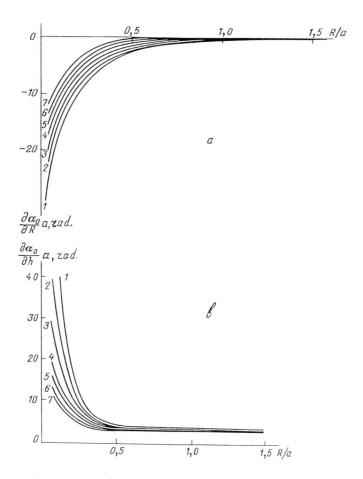

Fig. 2.12. $\partial\alpha_0/\partial R$ (a) and $\partial\alpha_0/\partial h$ (b) vs crystal radius for various growth angles ψ_0: (1) $-15°$, (2) $-5°$, (3) 0, (4) 10°, (5) 20°, (6) 30°, (7) 40°.

2.2.7. Capillary Stability

Let notation of capillary stability be introduced. Consider the case when a change in crystal dimensions and crystallization front position does not lead to a change in liquid- and solid-phase temperature gradients at the crystallization front. In this case, our system stability can be provided only by the capillary effects (capillary stability). In practice, a crystallization system of one degree-of-freedom is considered in this case and the set of equations (1.3) describing stability will reduce to the simple relation $\delta\dot{R} = A_{RR}\delta R$. The condition $A_{RR} < 0$ is a condition necessary and sufficient to attain capillary stability. In this case, any crystal dimension perturbation will attenuate.

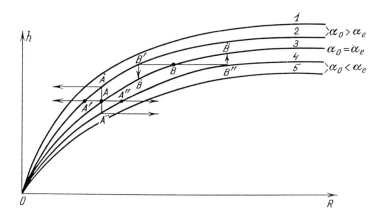

Fig. 2.13. Profile curve height vs. crystal radius for various growth angles ϕ_0: (1) $-10°$, (2) 0, (3) $10°$, (5) $30°$. A – scheme of capillary instability, B – scheme of thermal stability.

A positive value of the coefficient A_{RR} is indicative of capillary stability absence during crystal pulling by the Czochralski technique. This is illustrated in Figure 2.13. Let point A characterize constant-diameter crystal growth ($\alpha_0 = \alpha_e$). An arbitrary change in the crystal dimensions, upward tapering A', downward tapering A'', or in the crystallization-front position, rise $\dot{A}$, fall $\ddot{A}$, will cause such a change of the angle α_0 that it will lead to further crystal tapering according to Figure 1.8.

Crystallization stability can be ensured here only at the expense of the heat effects if the conditions (2.41) are satisfied. In this case, a crystal dimension change (points B' and B'', Figure 2.13) results in such crystallization front change (points $\dot{B}$ and $\ddot{B}$, Figure 2.13) that it provides growth of crystals of slightly changed dimensions with the liquid-solid phase conjugation angle equal to the growth angle.

2.3. The Heat Problem

The problem of applicability of the one-dimensional thermal-conductivity equation for temperature distribution description in the crystal-melt system and use of quasi-stationary approach to stability study was discussed in 1.6. In that case the stationary state of the crystal-melt system should be analyzed for stability. In practice, the stationary state cannot be achieved while growing crystals by the Czochralski technique as the crystal length increases and the melt level in the crucible decreases during the process of growth. Temperature distribution during crystal growth by the Czochralski technique was calculated in [88, 108, 109]. In all the calculations a crystal was supposed to have constant dimensions and be infinitely long, i.e., the stationary state was assumed to have been achieved.

As was estimated in [108], for germanium the crystal length that practically provides independence of temperature distribution on crystal length near the crystallization front is equal to $6\sqrt{R}$. Thus, for dielectrics and semiconductors possessing low thermal conductivities, stationary state is achieved rather quickly. It was for these materials that calculations were performed in [108, 109]. Thermal conductivities of metals are an order of magnitude higher, therefore near-stationary state cannot be achieved with the crystal length indicated [110] (high thermal conductivities allow application of the one-dimensional thermal-conductivity equation for calculation of temperature distribution in these crystals [110, 111]). However, in all these cases our analysis is applicable if the time interval of noticeable changes in the system parameters resulting from crystal growth is much longer than the time interval of the transient processes in the crystal-melt system. For practically used rates of crystal growth, this statement is valid, therefore any state of the system not subjected to other perturbances except for changes in the crystal length and melt level in the crucible caused by the growing process itself can be regarded as a quasi-stationary one.

2.3.1. *Problem Formulation*

Thus, the stationary state temperature distribution in the liquid and solid phases in the coordinate system shown in Figure 1.1 will be described by (1.40) with zero left side.

$$\frac{d^2 T_i}{dz^2} - \frac{V}{k_i}\frac{dT_i}{dz} - \frac{\mu_i}{\lambda_i}F(T_i - T_e) = 0 \tag{2.42}$$

If the true meniscus shape is referred to as a straight circular cylinder of the radius R as the first approximation while solving this equation for the liquid phase, the cross-section perimeter-to-its area ratio (F) both for the meniscus and for the crystal will be equal to $2/R$.

Now we shall define the boundary conditions of our problem. We shall specify the melt temperature at the bottom of the liquid column at the level of the melt-free surface:

$$T_L \big|_{z=0} = T_m \tag{2.43}$$

The crystallization-front temperature is fixed:

$$T_L \big|_{z=h} = T_S \big|_{z=h} = T_0 \tag{2.44}$$

For the boundary condition at the top of the crystal, a few versions practically applicable to various pulling schemes, will be considered.

A semi-infinite crystal, continuous pulling: the temperature at the crystal top is equal to the environment temperature:

$$T_S \big|_{z\to\infty} = T_e \tag{2.45}$$

A crystal of finite length, the temperature at the crystal top is specified:

$$T_S \big|_{z+1} = T_1 \tag{2.46}$$

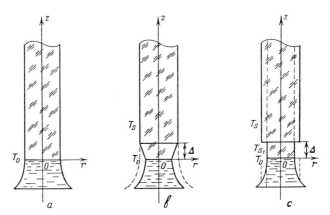

Fig. 2.14. Steady-state growth (a), real change of crystal dimensions (b) and an idealized diagram (c).

A crystal of finite length, heat removal from the crystal top is negligible:

$$\frac{dT_S}{dz}\Big|_{z=1} = 0 \tag{2.47}$$

We will not write out the solutions of the thermal conductivity equation as the expressions are rather lengthy and their derivation does not present any major difficulties [87]. Only the explicit form of the coefficients A_{hh} and A_{hR} obtained from the formulae (2.5), (2.6) as a result of solving the heat problem will be given.

2.3.2. Thermal Stability

The coefficient A_{hh} is associated with the crystallization-front displacement and indicates self-stabilization of the parameter h. Indeed, if the crystal radius is considered to be fixed (e.g., Bridgman growth) the sign of the coefficient A_{hh} determines crystallization stability. We can say that the system is thermally stable if $A_{hh} < 0$ (thermal stability by analogy with capillary stability).

The coefficient A_{hR} indicates interstabilization of the parameters, i.e., the crystallization front position and crystal or melt column dimensions. In this case the liquid column dimensions change all over its length, while the crystal dimensions change only near the crystallization front (Figure 2.14).

Accurate calculation of the contribution of crystal-dimension changes near the crystallization front to the coefficient A_{hR} is rather a lengthy problem, therefore only two limiting cases (Figure 2.14c) will be analyzed[2]: an allowance for crystal dimension changes all over its length and absence of contributions associated with

---

[2]In previous works [51, 91], we attempted to take this effect into account by introducing various functions $F(z)$ into (2.42). Equations are thereby obtained that can be solved using either the Weber or Airy functions. However, these results are not cited here as such an approach is not strict enough.

crystal dimension changes, the latter being easily obtained from the former if all the terms relating to the solid phase and depending upon $\partial/\partial R$ are equated to zero in the expressions for A_{hh} and A_{hR}.

2.3.3. *A semi-infinite crystal (heat removal from the crystal lateral surface is allowed for, while that from the melt is not taken into consideration: $\mu_S \neq 0; \mu_L = 0$; the boundary conditions are (2.43)–(2.45)).* This approximation is valid for long-crystal growth (the limiting case is continuous pulling), with good thermal screening of the melt column provided. For A_{hh} and A_{hR} we obtain:

$$A_{hh} = -\frac{\lambda_L}{\mathcal{L}} \frac{T_m - T_0}{h^2} \tag{2.48}$$

$$A_{hR} = \frac{\mu_S}{\mathcal{L}R^2} \frac{T_0 - T_e}{\zeta_S} > 0 \tag{2.49}$$

Here

$$\zeta_S = \sqrt{\frac{V^2}{4k_S^2} + \frac{2\mu_S}{\lambda_S R}}$$

In the case under consideration, the thermal conditions stabilize the crystallization front position ($A_{hh} < 0$) if the melt is overheated ($T_p > T_0$). Indeed, if an arbitrary perturbation leads to h increase, the liquid-phase temperature gradient decreases, and hence heat supply to the front decreases. In this case, the true crystallization rate should increase, which will result in downward front displacement, and hence, decrease in h.

The inequality $A_{hR} > 0$ can be explained in the following way. The heat flow, $\pi R^2 G_S$, removed via the crystal lateral surface is proportional to the lateral surface area, i.e., R. Therefore, with the crystal radius increasing, G_S decreases, which leads to h increasing.

As follows from the previous paragraph, for the Czochralski technique $A_{hR} < 0$ and hence $A_{Rh} A_{hR} < 0$, which contributes to satisfying (2.11). It therefore follows that interstabilization of the parameters is available in the system. Indeed, if the crystal diameter uncontrollably increases, the crystallization front height will increase due to $A_{hR} > 0$. This will result in α_0 increase as $\frac{\partial \alpha}{\partial h} 0 > 0$, and hence in crystal contraction according to the diagram given in Figure 1.8.

As interstabilization exists, stable crystal pulling by the Czochralski technique is possible if capillary stability is absent. To achieve this, it is necessary and sufficient that the following inequalities obtained from (2.41) should be satisfied:

$$\left|\frac{\partial \alpha_0}{\partial R}\right| < \frac{\lambda_L}{\mathcal{L}Vh^2}(T_m - T_0) \tag{2.50}$$

$$\left|\frac{\partial \alpha_0}{\partial h}\right| > \left|\frac{\partial \alpha_0}{\partial R}\right| \frac{T_m - T_0}{T_0 - T_e} \frac{R^2 \lambda_L \zeta_S}{h^2 \mu_S} \tag{2.51}$$

2.3.4. *Crystal of finite length, l, (the temperature of the crystal upper end is equal to the environmental temperature:* $\mu_L = 0, \mu_S \neq 0$, the boundary conditions are (2.43), (2.44), (2.46), $T_1 = T_e$). The approximation most of all fits growth of finite-length crystals with the seed temperature fixed, e.g., a water-cooled seed, and heat-screened melt column:

$$A_{hh} = -\frac{1}{\mathcal{L}}\left[\frac{\lambda_L(T_m - T_0)}{h^2} + \frac{\lambda_S\zeta_S^2(T_0 - T_e)}{sh^2[\zeta_S(l-h)]}\right] \tag{2.52}$$

$$A_{hR} = \frac{(T_m - T_e)\mu_S}{\mathcal{L}R^2\zeta_S}\left[cth[\zeta_S(l-h)] - \frac{\zeta_S(l-h)}{sh^2[\zeta_S(l-h)]}\right] \tag{2.53}$$

In principle, this case differs from the previous one in the fact that the solid-phase temperature gradient depends upon the crystallization front position, the gradient modulus increasing with h. This results in the second term appearing for A_{hh} in (2.52) which is negative, and hence improves the process stability. For small $(l-h)$ values, we can expand $sh^2[\zeta_S(l-h)]$ into a series restricting our analysis to the first term of the expansion and then the second term in (2.52) is obtained similar in structure to the first one: $\lambda_S(T_0-T_e)/(l-h)^2$. The coefficient A_{hR} in (2.53) differs from that in (2.49) in the factor within the parentheses that tends to one with l increasing. If l is small (approaches h in the order of magnitude), the factor tends to $2\zeta_S(l-h)/3$.

2.3.5. *Crystal of finite length l, (heat removal through the end face is absent:* $\mu_L = 0, \mu_S \neq 0$; the boundary conditions are (2.43), (2.44), (2.47)). The approximation is valid for crystals with heat-insulated end face and screened melt column:

$$A_{hh} = -\frac{\lambda_L}{\mathcal{L}}\frac{T_m - T_0}{h^2} + \frac{2\mu_S(T_0 - T_e)\zeta_S^2}{\mathcal{L}R\left[\frac{V}{2k_s}[\zeta_S(l-h)] + \zeta_S ch[\zeta_S(l-h)]\right]} \tag{2.54}$$

$$A_{hR} = \frac{\mu_S(T_0 - T_e)}{\mathcal{L}R^2\zeta_S} \times$$

$$\frac{\zeta_S(l-h)\frac{2\mu_S}{\lambda_S R} + \frac{1}{2}\left[\zeta_S^2 + \left(\frac{V}{2k_s}\right)^2\right]sh[2\zeta_S(l-h)] + \frac{2\zeta_S V}{2k_s}sh[\zeta_S(l-h)]}{\left[\frac{V}{2k_s}sh[\zeta_S(l-h)] + \zeta_S ch[\zeta_S(l-h)]\right]^2} \tag{2.55}$$

In the expression for A_{hh} positive additional terms that worsen stability appear. This is caused by the fact that heat removal from the crystal is provided only through its lateral surface, while the lateral surface area decreases with the crystallization front displacement in the upward direction and increases with the crystallization front displacement in the downward direction. If $l \gg h$, the value of the correction factor is small.

2.3.6. *Semi-infinite Crystal (heat removal from the melt-column lateral surface is allowed for:* $\mu_L \neq 0, \mu_S \neq 0$; the boundary conditions are (2.43)–(2.45)). If the melt-column screening is not effective enough to ignore heat removal from its surface, instead of the expressions (2.48) and (2.49) for A_{hh} and A_{hR} we obtain rather awkward expressions including hyperbolic functions. Regarding h as a small value, we can restrict our analysis to two first terms when expanding the hyperbolic functions into a series:

$$A_{hh} = -\frac{\lambda_L}{\mathcal{L}} \left[\frac{T_m - T_0}{h^2} + \frac{1}{2}(T_m - T_0)\zeta_L^2 \left(1 - \frac{Vh}{3k_L} \right) \right] \tag{2.56}$$

$$A_{hR} = \frac{1}{\mathcal{L}R^2} \left[\frac{(T_0 - T_e)\mu_S}{\zeta_S} - \frac{\mu_L(T_0 - T_e)(1 + Vh/2k_L)h}{3} \right.$$
$$\left. - \frac{\lambda_L(T_0 - T_e)\zeta_L h}{2} \right] \tag{2.57}$$

The coefficient A_{hh} in (2.56) is differs from that of (2.48). If not-overheated-melt pulling is carried out $(T_p = T_0)$, $A_{hh} = 0$ in (2.48), while in (2.56) $A_{hh} < 0$, e.g., the system stability is improved. This is attributed to the fact that the melt gets cooled while moving along the column, creating a temperature gradient. Changes in the melt-column height cause changes in the gradient value, thus stabilizing the system.

As compared with (2.49), the positive coefficient A_{hR} decreases (as was mentioned above, $A_{hR} > 0$ in (2.49) increases due to crystal surface cooling, while melt-column cooling results in an opposite effect).

When melt-column cooling is allowed for, quite similar results are obtained for a finite-length crystal with a specified temperature at its end, which is shown by additional terms appearing in (2.52), (2.53).

2.3.7. *The environmental temperature depends on the z-coordinate* $(T_e = T_e(z))$. Assume that the environmental temperature is linearly dependent on the z-coordinate: $T_e = T_{e0} - G_e z$. If in this case the reduced temperature $T_i^* = T_i + G_e z - G_e \lambda_i V / F k_i \mu_i$ is introduced into (2.42) instead of T_i, the form of the equation will not change.

In this case, a negative additional term will appear in all the expressions for the coefficient A_{hh}. Its value is mainly determined by the environmental temperature gradient G_e. Physically, it is easy to explain. With the crystallization front displacing towards the colder zone, heat radiation from the lateral surface increases, and the liquid- and solid-phase temperature gradients change in such a way that the crystallization front goes down. The opposite can be observed when the crystallization front goes down.

2.3.8. *Quantitative Stability Criteria and Quasi-Stationary Approach Applicability*

Using (2.50) and (2.51), the numerical values of the parameters ensuring stability of Czochralski growth can be found. When growing metal and semiconductor single crystals, it can be assumed that $V \approx (10^{-3}$–$10^{-4})$ m/s, $\mu_i \approx 10 \mathrm{Jm}^{-2}\mathrm{K}^{-1}$,

$\mathcal{L} \approx 5 \; 10^9 \mathrm{Jm}^{-3}$, $\lambda_i \approx 10^2 \; \mathrm{Jm}^{-1}\mathrm{s}^{-1}\mathrm{K}^{-1}$, $G_L \mid_{z=h} \approx (10^3-10^4) \; \mathrm{Km}^{-1}$, $h \approx 10^{-3}\mathrm{m}$, $R \approx 10^{-2}\mathrm{m}$, $T_m - T_0 \approx 10$ K, the heat capacity $c \approx 1$ J $\mathrm{kg}^{-1}\mathrm{K}^{-1}$, $a \approx 5 \; 10^{-3}\mathrm{m}$. Hence, it follows that for overheated-melt growth $A_{hR} \approx (0.1-1)\mathrm{s}^{-1}$, $A_{hh} \approx$ -(1–10)s^{-1} and on the basis of (2.50) and (2.51) stable-growth conditions for the Czochralski technique are obtained.

$$\left| \frac{\partial \alpha}{\partial R} 0 \right| a \leq 5, \qquad \frac{\partial \alpha}{\partial h} 0 \geq 10 \left| \frac{\partial \alpha}{\partial R} 0 \right|. \tag{2.58}$$

Comparison of (2.58) with the plots (Figure 2.12) leads to the conclusion that only for $R/a \geq 1$ are both the inequalities satisfied, i.e., stable growth of cylindrical crystals by the Czochralski technique is possible.

Now applicability of the quasi-stationary approach used to analyze the thermal conditions of the Czochralski crystallization technique will be verified.

The coefficient $A_{hh} = \delta \dot{h}/\delta h$ used in (1.46) is the quantity characterizing the crystallization front relaxation time:

$$1/\tau_{rf} = \mid A_{hh} \mid$$

Hereafter the following relations can be used to estimate of τ_{rf} and τ_{rT}:

$$\tau_{rf} \approx 1/ \mid A_{hh} \mid \approx \mathcal{L}h^2/\lambda_L(T_m - T_0), \tau_{rT} \approx h^2/k_L$$

These relations follow from (2.48), (1.47) and they change very little in order of magnitude when an allowance for the specific features of the heat problem is made. Now these relations will be rearranged:

$$\tau_{rf}/\tau_{rT} \approx \mathcal{L}/\rho c(T_m - T_0) \tag{2.59}$$

Here ρ is the density, c is the specific heat. When estimating these relations for the case of metal and semiconductor crystal growth, we obtain:

$$\tau_{rf}/\tau_{rT} \approx 10^2 \gg 1$$

The criterion of (1.46) holds, and the quasi-stationary approach is applicable.

2.3.9. *Allowance for Growth-Rate Dependence of the Crystallization Temperature*

In all the above calculations the crystallization temperature T_0 was supposed to be constant and equal to the equilibrium temperature of solid body-melt phase transition. However, in practice, the crystallization temperature differs from the equilibrium one and is a function of the crystallization rate: $T_0 = T_0(V)$. In other words, to obtain the crystallization rate specified, supercooling should be provided, $\delta T = T_{0e} - T_0(V)$, Thus, the values of the temperature gradients in (2.5), (2.6) depend not only on the instantaneous crystallization-front position but also on the instantaneous crystallization rate. This leads to the following changes in the form of (2.5) and (2.6) for A_{hR} and A_{hh} , respectively:

$$A_{hR} = \frac{\frac{\partial}{\partial R}(\lambda_S G_S - \lambda_L G_L)}{\mathcal{L} - \frac{\partial}{\partial T_0}(\lambda_S G_S - \lambda_L G_L)\frac{\partial T_0}{\partial V}} \tag{2.60}$$

$$A_{hh} = \frac{\frac{\partial}{\partial h}(\lambda_S G_S - \lambda_L G_L)}{\mathcal{L} - \frac{\partial}{\partial T_0}(\lambda_S G_S - \lambda_L G_L)\frac{\partial T_0}{\partial V}} \tag{2.61}$$

Based on the heat-problem solution for the crystal-melt system in the above-mentioned formulation: Equation (2.42), boundary conditions (2.43)–(2.45), the melt- and crystal-temperature gradients in their simplified forms can be represented as follows:

$$G_L \approx \frac{T_m - T_0}{h}, \qquad G_S \approx \zeta_S(T_0 - T_e) \tag{2.62}$$

In this case the denominator of (2.60) and (2.61) will have the form:

$$\mathcal{L} - \left(\lambda_S \zeta_S + \frac{\lambda_L}{h}\right)\frac{\partial T}{\partial V^0} \tag{2.63}$$

Generally, $\partial T_0/\partial V < 0$, therefore $|A_{hh}|$ decreases and the front relaxation time increases.

Indeed, if the crystallization front height increases, the module of the melt temperature gradient decreases. To maintain heat balance, the crystallization front begins to move down, with the crystallization rate increasing. This rate increase, in its turn, causes a decrease in the crystallization temperature and hence an increase of the modulus of the melt temperature gradients. The real crystallization rate decreases and the crystallization front returns to the equilibrium position more slowly. For various growth mechanisms the following relations between supercooling and the growth rate hold: linear, quadratic and exponential depending on the facet growth mechanisms, normal, dislocation or singular-facet ste-forming ones, respectively [83].

For linear and quadratic relations the value of $\frac{\partial T}{\partial V}0 \sim \frac{\delta T}{V}0$ and hence the correction to $\mathcal{L}$ in (2.63) is of the order of magnitude of $\lambda_L(T_p - T_0)V^{-1}h^{-1}$ and becomes appreciable only in the case when the heat flow $\lambda_L \delta T/h$ caused by crystallization front supercooling is comparable with the crystallization heat flow $\mathcal{L}V$.

For slight melt supercooling $(T_p \leq T_{0e})$ an allowance for the crystallization-front kinetic effects [112] leads to true supercooling of the melt $[T_p > T_0(V)]$ and heat stability existence in the system $(A_{hh} < 0)$.

2.4. Crystallization Process Stability

The method of crystallization stability analysis described previously is a formalized one. Now a more clear approach from the physical point of view will be applied. Assume that a crystal of some specified diameter is to be grown by the Czochralski technique. With the growth angle specified, the crystallization front position, point A on the curve $h(R)\mid_{\alpha=const}$ (Curve 1, Figure 2.15) can be calculated by

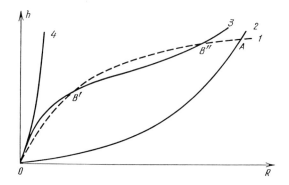

Fig. 2.15. The functions $h(R)$ obtained by solving: 1 – the capillary problem; 2, 3, 4 – the heat problem for various crystallization parameters. The points A, B' and B'' represent simultaneous capillary- and heat-problem solutions.

solving the capillary problem. The crystallization front position h can also be calculated by solving the heat problem for the crystal-melt system, point A on the curve $h(r)\,|_{V=const}$ (Curve 2, Figure 2.15).

Thus, the h-value must simultaneously be the solution of both the capillary and the heat problems. Then the problems of uniqueness and stability of the simultaneous capillary- and heat-problem solution arise. Diagrams similar to that given in Figure 2.15 can help to study these problems.

Independently, the curve $h(R)$ obtained by solving the capillary problem can be plotted. Furthermore, series of the curves $h(R)$ determined by solving the heat problem for the melt-crystal system can be plotted. In these cases either the pulling rate or the melt temperature, or some other parameter can be used as a variable. For some values of these parameters, simultaneous solutions do not exist (Curve 4, Figure 2.15). If simultaneous solutions exist, their stability can be analyzed: the conditions when (2.10), (2.11) hold are found. This will be done for the techniques of pulling from shaper in the next chapter.

Based on the analysis carried out, the main data on the stability of crystallization by the Czochralski technique obtained here can be summarized. The Czochralski technique is referred to as a system with two degrees-of-freedom.

1. *Capillary stability is absent: $A_{RR} > 0$.*

2. *Interstabilization of the parameters is observed: $A_{Rh} A_{hR} < 0$.*

3. *As a result, stable growth is possible only in case heat stability is provided $A_{hh} < 0$, and the condition $A_{hh} > A_{RR}$ is satisfied.*

4. *The condition $A_{hh} < 0$ is easily satisfied by providing overheated-melt growth.*

5. *The condition $A_{hh} > A_{RR}$ is easily satisfied by providing growth of crystals of large diameters, $R > a$, for which $A_{RR} \to 0$.*

6. *Providing process stability by increasing overheating is associated with the difficulties of heat removal from the crystallization front. Therefore, a technique increasing the coefficient of convective heat transfer from the crystal surface, e.g., by blowing the crystal localized zone above the crystallization front with gas, can be recommended.*

Some of these conditions of stability were actually realized in [113] when crystals of constant cross-section were grown.

Chapter 3

TECHNIQUES OF PULLING FROM SHAPER

While analyzing the stability of techniques of pulling from shaper (TPS) as a system with two degrees-of-freedom, the data of 2.1 can be used. The set of Equations (2.1), (2.2), the expression for the coefficients (2.3)–(2.6) and inequalities (2.10), (2.11) characterizing the stability conditions will be the same for the Czochralski technique and for TPS. Therefore we can proceed to the analysis of the melt-column shaping conditions in TPS.

3.1. Boundary Conditions of the Capillary Problem

As was mentioned above, when formulating the boundary problem for the Laplace capillary equation, two boundary conditions should be specified. As follows from the description of the pulling techniques, one of them will be specified at the lower boundary of the melt column, the other one being specified at the upper boundary, i.e., along the liquid-crystal contact line. A shaper is used for melt-column shaping at its lower boundary in TPS (Figure 3.1). The functions of the shaper in TPS are varied. Here the shaper is referred to as a device to control the melt-column shape [29–31] only. Shaping is accomplished either on the walls (Figure 3.1b, c) or on the sharp edges (Figure 3.1a) of the shaper.

The shaper (Figure 3.1) will be characterized by its depth d_0, its wall or free edge-curvature radius r_0 in the horizontal plane, the angle B_0 made by its wall with the horizontal. The wetting angle θ formed by the melt and the shaper wall is a very important shaper characteristic. If this angle exceeds $90°$, the shaper material is not wetted by the melt. If it is smaller, the melt wets the shaper material.

The following boundary conditions on the shaper walls and at its free edges are possible.

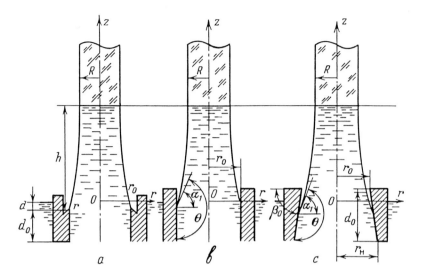

Fig. 3.1. Boundary conditions specified by the shaper: the catching condition (a), wetting on the vertical (b) and inclined (c) walls. The notations are in the text.

3.1.1. Catching Conditions

In case the shaper material is wetted by the melt, the liquid is easily caught by its sharp edge. As was mentioned above, this boundary condition will be called the *catching condition*. For sharp shaper contours Γ this condition will take the following form (Figure 3.1a):

$$z \mid_\Gamma = d(r, \varphi) \tag{3.1}$$

Here $d-$ is the distance from the shaper edge to the coordinate plane which coincides with the liquid surface plane. Where the shaper is flat and is positioned parallel to the liquid plane, $d = const$ and represents the pressure of feeding the melt to the shaper. The pressure being positive, the shaper is positioned below the melt free-surface level and vice versa.

If the shaper is wettable, its free edges should really be sharp (Figure 3.1a) so that the melt does not spread over the shaper surface and it can function properly.

The possibility of providing the catching conditions at non-wettable shaper free-edges will be discussed below.

3.1.2. Angle Fixation Condition (The Wetting Condition)

If the liquid does not reach the sharp shaper edges, its surface makes the wetting angle with the shaper walls (Figure 3.1b, c). As was shown above, this boundary condition in a general case will be called *the angle fixation condition*, and for the shaper it will be called *the wetting condition*. It can be written down in the following form [65]:

$$\tilde{H}\frac{\partial z}{\partial \mathbf{n}}|_{\Gamma} = -\cos\theta \tag{3.2}$$

Here $\tilde{H}$ is the same as in (1.10), $\mathbf{n}-$ denotes the direction of the inner normal towards the shaper wall (Figure 3.1c). It should be noted that where this boundary condition is satisfied, the angle is fixed on the shaper wall. In this case the wall itself can be wetted ($\theta < 90°$) or not wetted ($\theta > 90°$) by the melt.

3.1.3. *Wetting-To-Catching Condition Transition*

With the melt pressure increasing, the catching condition at the non-wettable shaper free-edges can be obtained. Figure 3.2a illustrates this transition. The diagram is based on [107] where the particle buoyancy conditions for the flotation processes are described. The relations of the transition mentioned will be obtained in 3.8. The pressure will be increased by gradual shaper immersion into the liquid. A nonwettable shaper with a hole possessing vertical walls will be considered and a number of its successive positions will be analyzed.

Position A: the shaper touches the melt with its lower plane. The lower-plane immersion depth is equal to zero. The angle χ between the shaper wall and the melt plane is 90°. The catching condition holds at the lower free edge of the shaper. With the shaper being immersed into the liquid, the angle χ increases, and when the shaper lower plane reaches some depth d_1, (position B) the angle χ will be equal to the wetting angle θ, (within the immersion depth range from 0 to d_1 (the catching condition holds at the lower free edge). With further shaper immersion into the liquid, the angle χ between the line tangent to the liquid surface and the shaper wall remains equal to the wetting angle θ. The liquid-shaper wall contact line goes up by the value of l^1, the shaper immersion depth, the distance between this line and the free surface level remaining equal to d_1 (position C).

As soon as the immersion depth is equal to $d_0 + d_1$, the liquid-shaper contact line will coincide with the shaper sharp edge (position D) and with further pressure increase the catching condition at the shaper upper free edge will hold (position E). In this case the angle χ will increase until the wetting angle θ is formed by the liquid and the shaper surface [5]. For the horizontal shaper surface $\chi = \pi/2 + \theta$ (position F). Further increase in pressure is impossible as it will lead to liquid spreading over the shaper surface and the shaper will not operate properly. But if shaper free edges are sharp, pressure increase is possible (Figure 3.1).

The presence of a seed or profile being pulled introduces considerable changes into the conditions of the transition described. It means that the condition at the upper boundary (along the crystal-melt contact line) can affect the character of the condition at the lower boundary (along the melt-shaper contact line). A diagram, that will be proved when solving the boundary problem, is given in Figure 3.2b. By changing the seed-to-shaper hole dimension ratio alone, the catching boundary condition at the

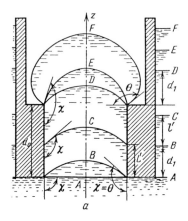

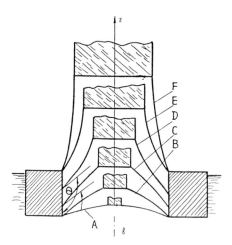

Fig. 3.2.   Transition of the shaper lower free-edge catching boundary condition (meniscus A) to the wetting condition (meniscus B, C, D) and further to the shaper upper free-edge catching condition (meniscus E, F). Pressure changes (the horizontal lines denote successive positions of the liquid free surface) (a); changes of the seed-shaper dimension ratio (b); ξ is the angle between the line tangent to the melt surface and the shaper wall.

lower free edge (meniscus A, B), the wetting condition on the shaper walls (meniscus C, D), the catching boundary condition at the upper free edge (meniscus E, F) can be achieved.

Hereafter, it will be shown that the catching boundary condition usually leads to capillary stability of the process, therefore, the ways of achieving the catching condition at the shaper free edges in TPS should be specified.

Firstly, a melt-wettable material should be used for the shaper, and the latter should be designed in such a way that the melt could rise up to the shaper free-edges due to capillary forces (Figure 1.3d, e, i, j).

Secondly, a melt-wettable material should be used, and the melt column should be supported from inside (Figure 1.3c).

Thirdly, for melt-nonwettable materials, the melt column should be embraced from outside providing additional pressure on the liquid to make the melt-shaper contact point touch the shaper sharp edge (Figure 1.3b, 3.2a).

Fourthly, for poorly wettable shaper materials, a seed of corresponding dimensions should be used in order to ensure melt column contact with the shaper sharp edges (Figure 3.2b).

3.1.4. *Influence of the Wetting-Angle Hysteresis on the Capillary Effects*

While analyzing all the capillary effects, the existence of wetting-angle hysteresis should be taken into consideration. The wetting angle hysteresis reveals itself in the fact [107] that the wetting angle of liquid run on a solid body θ_{on} is larger than that of liquid run off a solid body θ_{of}. This means that the stationary wetting angle depends upon the process of meniscus formation (on-run or off-run). This results, for example, in the fact that a higher pressure needs to be applied to create the catching condition at the shaper free edge than that required to keep it unchanged. Where the catching condition is created by the seed, this condition can remain unchanged in the process of growth, with the clearance between the shaper free edge and the growing crystal changing.

3.1.5. *Boundary Conditions with the Meniscus Axial Symmetry Being Available*

For cylindrical and flat profiles, conditions (3.1) and (3.2) will respectively assume the following forms (Figure 3.1a,b):

$$z \mid_{r=r_0} = d \tag{3.3}$$

$$z' \mid_{r=r_0} = -tg(\theta - \frac{\pi}{2}) = -tg\alpha_1 \tag{3.4}$$

Here, the case of an inclined-walled shaper should be considered separately (Figure 3.1c). Such a shaper is often used to improve the heat-exchange conditions between the crystal being pulled and the melt mass. The form of condition (3.3) for such shapers does not change, and condition (3.4) will be different:

$$z' \mid_{r=r_0 - ctbB_0z} = tg(\theta + B_0 - 180°) = -tg\alpha_1 \tag{3.5}$$

i.e., with the wetting boundary condition satisfied, the angle α_1 between the horizontal and the line tangent to the liquid surface at the point of its contact with the shaper wall can be changed by changing the angle between the shaper free edges B_0.

3.1.6. *Boundary Conditions at the Crystal Free Edges*

As previously been mentioned, while pulling a crystal of constant cross-section the angle fixation condition (2.14) should be satisfied over the whole contour $\hat{\mathcal{L}}$ (Figure 2.2c). In this case, the $\hat{\mathcal{L}}$-contour position relative to the shaper free edges or to the melt surface can be calculated by solving the boundary problem.

When meniscus catching on the crystal edge occurs, the $\hat{\mathcal{L}}$-contour position relative to the shaper free edges or the melt surface (2.15) can be specified (Figure 2.2). Then, as a result of the boundary problem solution, the angle made by the line tangent to the melt along the contour $\hat{\mathcal{L}}$ with any fixed direction can be calculated.

From now on, these boundary conditions will be specified according to a particular problem.

3.1.7. *Laplace Capillary Equation Form Depending on the Type of the Boundary Conditions*

To complete our analysis of the boundary problem formulation for the capillary equation, an important point needs to be stressed. The Laplace equation (1.12)–(1.15) can be written down with the origin of vertical coordinates selected at various points. Depending on this, d-parameter value will change. In this case, when the catching condition is satisfied at the shaper free edge, the equation can be written with either the free-surface level or the shaper level taken as the vertical-coordinate origin. In the former case, the pressure is included in the boundary condition, in the latter case it is included in the equation as a parameter.

In case the wetting condition is satisfied on the shaper wall, the shaper wall-melt contact point is not fixed, the pressure is not known and the Laplace equation can reasonably be written down taking the melt free-surface level alone as the vertical-coordinate origin. Then the pressure can be calculated by solving the boundary problem.

Availability of the liquid-surface differential equation in a convenient coordinate system with the parameters included in it and the boundary conditions strictly de-fined allows the formulation of a number of boundary problems. Their solution defines the limits of process parameter variations that provide meniscus existence, meniscus shape and explicit form of the coefficients of (1.39) for investigation of the crystallization-process capillary stability.

3.2. Review of Publications on Capillary Shaping in TPS

There were a few publications that preceded our works on the capillary-shaping conditions in TPS that should be mentioned.

In [68] profile curves for internal and external meniscus for pulling a circular tube are plotted by the Kelvin graphical method [102]. The case when the pressure is equal to zero is analyzed. For small-diameter tube growth, the Laplace equation ignoring

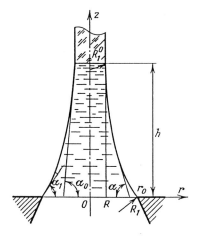

Fig. 3.3. Liquid column parameters for Equation (3.7).

the gravity pressure is analytically solved. A profile of constant cross-section is supposed to grow if the liquid-solid phase conjugation angle is equal to zero.

In [114], while crystallizing an aluminum tape the formula for the height h of the phase-transition plane as a function of the clearance l_0 between the shaper free edge and the tape pulled was experimentally checked:

$$h = \sqrt[3]{\frac{3}{4}a^2 l_0}. \tag{3.6}$$

This formula is obtained as a result of approximate integration of the Laplace equation if one of the main curvature radii is ignored and it makes sense for $h \leq 0.6a$. The disparity between the experimental results and the calculated ones observed is attributed to an oxide film formed on the aluminum surface.

In a number of works, for instance [115], to calculate the crystallization front height it is proposed to use the formula obtained in [71] on the basis of the approximation of (2.23)–(2.24) considered above. This formula has the following form in our notations, shown in Figure 3.3.

$$\frac{h}{a} = \pm \left(\pm (\cos \alpha_0 - \cos \alpha_1) + \left[\frac{(A+1)\cos \alpha}{4RA} 0 - d \right]^2 \right)^{1/2}$$
$$- \left[\frac{(A+1)\cos \alpha}{4RA} 0 - d \right], \tag{3.7}$$

where $A = R_1 / R_1^0$.

However, application of formula (3.7) to TPS is impossible because of the following reasons [116]:

Firstly, when calculating the melt column height from formula (3.7), one should know the angles α_0 and α_1, which is equivalent to specifying the derivative dz/dr at the crystallization front and the shaper free edge. Furthermore, specifying the pressure is equivalent to specifying the coordinate of the melt column-base point. Thus, three boundary conditions instead of two are to be specified for the second-order differential equation. In practice, these conditions cannot be independent. For example, if the pressure d and the angle $\alpha_0 \mid_{r=R}$ are specified, the angle $\alpha_1 \mid_{r=r_0}$ should be calculated by solving the boundary problem, and vice versa, only one value of d corresponds to the specified angles $\alpha_0 \mid_{r=R}$ and $\alpha_1 \mid_{r=r_0}$.

Secondly, as the formula is applied to either concave or convex melt columns, a wide family of convex-concave melt columns for which the coordinates of the inflection points are unknown beforehand are excluded from consideration.

The possibility of applying the approximation from [71] to the Czochralski technique can be accounted for by the fact that the following three boundary conditions are nominally specified for the Czochralski technique:

$$\alpha \mid_{r=R} = \alpha_0, \quad z \mid_{r \to \infty} = 0, \quad \alpha \mid_{r \to \infty} = 0$$

This specific feature is associated with the fact that the conditions at the melt-column lower end are specified for $r \to \infty$, where with the z-coordinate tending to zero, the derivative dz/dr also tends to zero.

3.3. Liquid Meniscus Shapes

In the process of TPS growth, melt columns of three types can be formed: concave, convex and convex-concave ones (Figure 3.4). Analysis of the Laplace equation in the form of (1.12) allows the separation of the conditions of concave melt column existence on the one hand and those of convex and convex-concave ones on the other hand [29–31].

Indeed, if the second derivative $d^2z/dr^2 \mid_{r=r_0}$ at the free edge or on the wall of the shaper is positive, the profile curve is concave; if it is negative, the profile curve is either convex or convex-concave. From the profile curve equation (1.12) for the point $r = r_0$ we obtain the following relation:

$$z_0'' = -r_0^{-1}(1 + z_0'^2)[z_0' + 2r_0d(1 + z_0'^2)]^{1/2} \tag{3.8}$$

Here z_0' and z_0'' are the values of the first and second derivatives dz/dr and d^2z/dr^2 at the point $r = r_0$. The point r_0 is the inflection point if $z_0'' = 0$, i.e., if

$$-2r_0d = \frac{z_0'}{\sqrt{1 + z_0'^2}} \tag{3.9}$$

Since $z_0' = -tg\alpha_1$ (Figure 3.4), condition (3.9) can be rewritten in the following way:

$$\alpha_1 = \arcsin(2r_0d) \tag{3.10}$$

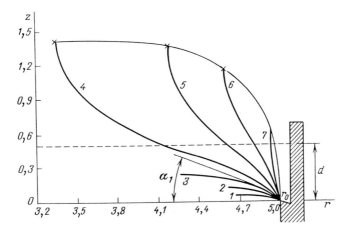

Fig. 3.4.   Convex (1, 2, 3), convexo-concave (4, 5, 6) and concave (7) profile curves plotted by computer-aided calculation of Equation (1.12) for $r_0 = 5.0$, $d = 0.5$. Crosses mark the profile-curve heights calculated from Equation (1.13).

Hence, the profile curve is concave if (Figure 3.5)

$$\alpha_1 \geq \arcsin(2r_0 d) \tag{3.11}$$

The profile curve is either convex or convex-concave if (Figure 3.5)

$$\alpha_1 \leq \arcsin(2r_0 d) \tag{3.12}$$

The range of convex and convex-concave curves existing will be defined below and the possible existence of curves of other types, e.g. concave-convex ones will be discussed.

3.4. Capillary Problem for Systems with Large Bond Numbers [31, 116]

Equation (1.13) describes the shape of the liquid column formed when pulling a crystal in the form of a plate or a rod of a large diameter as compared with the capillary constant, which corresponds to a large Bond number. If the vertical coordinate origin is chosen on the melt free-surface level, Equation (1.13) will assume the following form:

$$z'' - 2z(1 + z'^2)^{3/2} = 0. \tag{3.13}$$

The profile curve that is the solution of (3.13) is convex for $z < 0$, and concave for $z < 0$, as follows from the analysis of the sign of the second derivative in (3.13). Hence, the inflection point is on the level of the melt free surface (Figure 3.6).

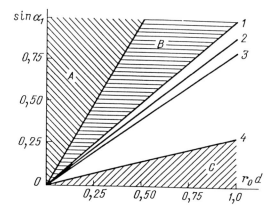

Fig. 3.5. Ranges of existence of concave (A), convexo-concave (B) and convex (C) profile curves. Lines 1, 2, 3, 4 separate ranges B and C for various r_0 values: 1 – corresponds to $r_0 = 1$; 2 – $r_0 = 0.5$; 3 – $r_0 = 1.0$; 4 – $r_0 = 5.0$.

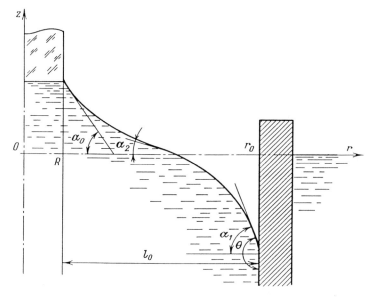

Fig. 3.6. Meniscus parameters (the angle-fixation boundary condition is specified at the crystal free edge, the wetting boundary condition being assigned on the shaper wall.

3.4.1. Angle-Fixation Boundary Conditions at Both Ends of the r-Variation Interval

The first boundary problem will be formulated using the following boundary conditions (Figure 3.6). The shaper wall is situated to the right where $r = r_0$. The line tangent to the melt surface in the liquid-shaper contact point and the negative direction of the r-axis make the angle α_1. Here $\alpha_1 = \theta - \pi/2(\theta$ is the angle of shaper wall wetting by the melt). The edge of a melt-grown flat crystalline tape or that of a circular cylindrical crystal of a large diameter is situated to the left where $r = R$. Let l_0 denote the clearance between the crystal edge and the shaper: $l_0 = r_0 - R$. Let the angle α_0 made by the line tangent to the melt surface at the melt/crystal grown contact point and the negative direction of the r-axis be specified (while growing crystals of constant cross-sections, the angle $\alpha_0 = \alpha_e = \pi/2 - \psi_0$ is the complement of the growth angle). Let us consider what process parameter data can be obtained with such problem formulation. The boundary conditions will be written down as follows:

$$\frac{dz}{dr}\Big|_{r=r_0} = -tg\alpha_1, \quad \frac{dz}{dr}\Big|_{r=R} = -tg\alpha_0 \tag{3.14}$$

According to our terminology, the angle-fixation boundary conditions on both ends of the r-variation interval can be obtained.

Profile curves that have positive and negative branches within the (R, r_0) interval will be sought. So, the first integral of (3.13) is obtained:

$$(1 + z'^2)^{-1/2} = A - z^2 \tag{3.15}$$

Here A is an arbitrary constant.

The limiting values of z can be found:

$$z(R) = (A - \cos\alpha_0)^{1/2}, \; z(r_0) = -(A - \cos\alpha_1)^{1/2} \tag{3.16}$$

The signs of the root are chosen proceeding from the physical sense of the quantities $z(R)$ and $z(r_0)$. Using the boundary conditions (3.14), the equation for A can be obtained:

$$\int_0^{\sqrt{A-\cos\alpha_1}} \frac{A - z^2}{\sqrt{1 - (A - z^2)^2}} dz + \int_0^{\sqrt{A-\cos\alpha_0}} \frac{A - z^2}{\sqrt{1 - (A - z^2)^2}} dz = l_0 \tag{3.17}$$

Assuming that $z = (A - \cos\alpha)^{1/2}$, this equation can be written down in the following form:

$$\int_{\alpha_2}^{\alpha_1} \frac{\cos\alpha}{(\cos\alpha_2 - \cos\alpha)^{1/2}} d\alpha + \int_{\alpha_2}^{\alpha_0} \frac{\cos\alpha}{(\cos\alpha_2 - \cos\alpha)^{1/2}} d\alpha = 2l_0. \tag{3.18}$$

Here $\alpha_2 = \arccos A$. As follows from (3.15), for meniscus having both positive and negative branches, α_2 is the angle between the line tangent to the meniscus and the r-axis on the level of the liquid free surface, i.e., at the inflection point (Figure 3.6). The left side of (3.18) is a decreasing function of the parameter α_2. Hence, its minimum value exists, let m denote it. This value is achieved when α_2 becomes equal to the smaller of the angles $\{\alpha_0; \alpha_1\}$. Only with $l_0 > m$, will the meniscus lie both above and below the liquid free surface. With $l \leq m$, the meniscus will lie either above or below the liquid free surface (Figure 3.7). The following notation will be introduced

$$\chi_1 = \min\{\alpha_1; \alpha_0\}, \chi_2 = \max\{\alpha_1; \alpha_0\}.$$

Then for m

$$2m = \int_{chi_1}^{\chi_2} \frac{\cos \alpha}{\cos \chi_1 - \cos \alpha)^{1/2} d\alpha}.$$

By elliptical substitution of the variables $sn^2(z, k) = (A + 1)^{-1}[1 + \cos \theta]$, where $k^2 = 1/2(A + 1)$ (sn is the Jacobi elliptical function, k is the Legendre modulus), the integral can be calculated [117]:

$$\int_{\varphi}^{\pi} \frac{\cos \vartheta}{\sqrt{A - \cos \vartheta}} d\vartheta = \sqrt{2}[F(k, \varphi) - 2E(k, \varphi)] \equiv G[k, \varphi].$$

Here F and E are the Legendre elliptical functions of the first and second kinds; they are very well tabulated in [117].

The amplitude is equal to

$$\varphi(\alpha, A) = \arcsin \sqrt{\frac{1 + \cos \alpha}{1 + A}}; \quad \text{here} \quad \alpha \geq \arccos A = \alpha_2.$$

Now (3.17) can be expressed in terms of the elliptical functions:

$$2G[k, \pi/2] - G[k, \varphi_1] - G[k, \varphi_0] = 2l_0. \tag{3.19}$$

Here

$$\varphi_1 = \arcsin \sqrt{\frac{1 + \cos \alpha_1}{1 + A}}, \quad \varphi_0 = \arcsin \sqrt{\frac{1 + \cos \alpha_0}{1 + A}}.$$

The minimum value of the left side of (3.18) as a function of α_2 is equal to

$$2m = G[k_1, \pi/2] - G[k_1, \varphi_{12}] \tag{3.20}$$

Where

$$k_1^2 = \frac{1 + \cos \chi_1}{2}, \quad \varphi_{12}(\chi_2, \chi_1) = \arcsin \sqrt{\frac{1 + \cos \chi_2}{1 + \cos \chi_1}}.$$

If χ_2 corresponds to the left end of the r-variation interval (to the crystal being pulled), with $l_0 < m$, the whole meniscus lies above the liquid free surface (Figure 3.7b); if χ_2 corresponds to the right end of the interval (to the shaper), the whole meniscus lies below the liquid free surface (Figure 3.7a). It is as if the larger angle overpulled the meniscus to its side. Note that this effect analytically obtained by us, was first pointed out by Poincaret [155].

$l_0 > m$. The accurate solutions expressed in terms of the function $G(k, \varphi)$ for meniscus parts located above and below the melt free-surface level can be written down in the following way:

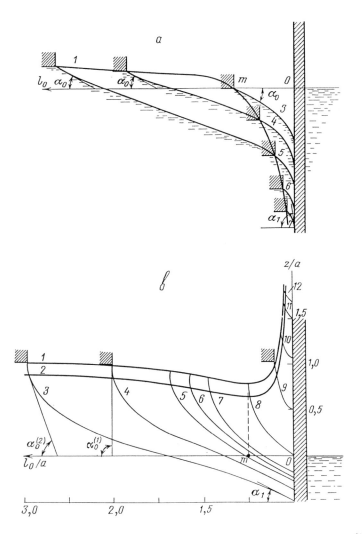

Fig. 3.7. Boundary (1, 2) and profile (3–12) curves: $\alpha_0 < \alpha_1$ (a); $\alpha_0 > \alpha_1$ with $\alpha_1 = 45°$, $\alpha_0^{(1)} = 90°$, $\alpha_0^{(2)} = 75°$ (b).

$$r(z) = R + 0.5\{G(k, \varphi_z) - G(k, \varphi_0)\}, \quad z \geq 0$$
$$r(z) = R + G(k, \pi/2) - 0.5\{G(k, \varphi_0) + G(k, \varphi_z)\}, \quad z \leq 0. \tag{3.21}$$

Here

$$\varphi_z = \arcsin \sqrt{1 - z^2(1 + A)^{-1}}.$$

$l_0 < m$. In this case (3.17) will have the following form:

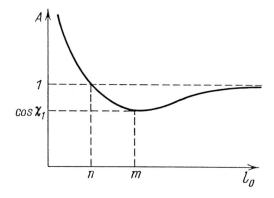

Fig. 3.8.   Integration constant A vs. crystal-shaper clearance value l_0.

$$l_0 = \int\limits_{\sqrt{A-\cos\xi_1}}^{\sqrt{A-\cos\xi_2}} \frac{A - z^2}{\sqrt{1 - (A - z^2)^2}} dz. \qquad (3.22)$$

Here A does not have such a clear physical sense as in the previous case, and no restrictions are imposed on its value; l_0 is a decreasing function of the parameter A (Figure 3.8).

The minimum A-value is equal to $\cos\chi_1$. With A changing from $\cos\chi_1$ to *one*, l_0 changes from m to n (Figure 3.8). Then l_0 can be represented in the following form:

$$l_0 = \int\limits_{\sqrt{A-\cos\chi_1}}^{\sqrt{A-\cos\chi_2}} \frac{A - z^2}{\sqrt{-(z^2 - \lambda^2)(z^2 - \mu^2)}} dz. \qquad (3.23)$$

Here $\lambda^2 = 1 - A$, $\mu^2 = 1 + A$. As in the previous case, to express this equation in terms of a set of the Legendre elliptical functions of the first and second kinds, the following substitution of variables should be made: $z^2 = \mu^2(1 - x^2)$. Then

$$2l_0 = G(k, \varphi_3) - G(k, \varphi_4).$$

Here

$$\varphi_3 = \arcsin\sqrt{\frac{1 + \cos\chi_1}{1 + A}}, \qquad \varphi_4 = \arcsin\sqrt{\frac{1 + \cos\chi_2}{1 + A}}$$

$A = 1.0$.   In this case the integral (3.22) can be expressed in terms of elementary functions:

$$n \equiv l_0 \mid_{A=1} = \frac{1}{\sqrt{2}} \ln\frac{\sqrt{2} + \sqrt{1 + \cos\chi_1}}{\sqrt{1 - \cos\chi_1}} -$$

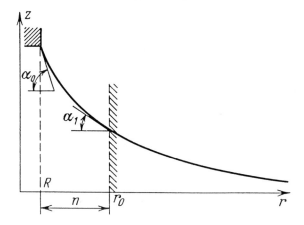

Fig. 3.9. The profile curve for $A = 1$.

$$\frac{1}{\sqrt{2}}ln\frac{\sqrt{2}+\sqrt{1+\cos\chi_2}}{\sqrt{1-\cos\chi_2}}+\sqrt{1+\cos\chi_2}-\sqrt{1+\cos\chi_1} \tag{3.24}$$

Then the profile curve is part of the meniscus obtained without a shaper and enclosed between the points R and r_0 at which the derivatives are equal to $tg\alpha_1$ and $tg\alpha_0$ (Figure 3.9).

$A > 1$. In this case l_0 is specified by (3.23) if

$$\lambda^2 = A - 1, \mu^2 = A + 1 \tag{3.25}$$

Substation of the variables can be made to pass to the elliptical functions

$$z^2 = \mu^2 - (\mu^2 - \lambda^2)x^2$$

Then

$$l_0 = \frac{A}{\sqrt{A+1}}[F(k_2, \varphi_5) - F(k_2, \varphi_6)]$$
$$-\sqrt{A+1}[E(k_2, \varphi_5) - E(k_2, \varphi_6)]. \tag{3.26}$$

Here

$$\varphi_5 = \arcsin\sqrt{\frac{1+\cos\chi_2}{2}}, \quad \varphi_6 = \arcsin\sqrt{\frac{1+\cos\chi_1}{2}}.$$
$$k_2^2 = \frac{\mu^2 - \lambda^2}{\mu^2} = \frac{2}{A+1}.$$

The profile-curve equation in terms of the functions F and E with $l_0 < m$ can be written as follows.

In case $A < 1(m > l_0 > n)$:

$$r(z) = l_0 - \frac{1}{2}[G(k, \varphi_3) - G(k, \varphi_z)] \tag{3.27}$$

In case $A < 1 (l < n)$:

$$r(z) = l_0 - \frac{A}{\sqrt{A+1}}[F(k_2, \varphi_{z1}) - F(k_2, \varphi_6)]$$

$$+\sqrt{A+1}[E(k_2, \varphi_{z1}) - E(k_2, \varphi_6)] \tag{3.28}$$

Here

$$\varphi = \arcsin\sqrt{\frac{1 - A - z^2}{2}}.$$

For the limiting case when $l_0 \to 0$ by substitution the variables $z = \sqrt{A - \cos\alpha}$ gives the following form of the equation for l_0:

$$l_0 = \int_{\chi_1}^{\chi_2} \frac{\cos\alpha}{\sqrt{A - \cos\alpha}} d\alpha.$$

With $l_0 \to 0$, A tends to infinity, hence

$$l_0 \approx A^{-1/2} \int_{\chi_1}^{\chi_2} \cos\alpha \, d\alpha \approx A^{-1/2}(\sin\chi_2 - \sin\chi_1) \tag{3.29}$$

Here $A = (\sin\chi_2 - \sin\chi_1)^2 / l_0^2$. From here, the maximum height of liquid rise (lowering) is equal to

$$z_{max} = \sqrt{\frac{(\sin\chi_2 - \sin\chi_1)^2}{l_0^2} - \cos\chi_1}. \tag{3.30}$$

The minimum height is equal to

$$z_{min} = \sqrt{\frac{(\sin\chi_2 - \sin\chi_1)^2}{l_0^2} - \cos\chi_2}. \tag{3.31}$$

Example. Consider the way a profile curve is plotted using the results obtained. Let $\alpha_1 = 45°$, $\alpha_0 = 75°$ and in accordance with our choice $\chi_1 = 45°$, $\chi_2 = 75°$ (Figure 3.7b).

(a)   Firstly, m should be calculated:

$$2m = G\left(\frac{\pi}{2}; \sqrt{\frac{1 + \cos 45°}{2}}\right)$$

$$-G\left(\arcsin\sqrt{\frac{1 + \cos 75°}{1 + \cos 45°}}; \sqrt{\frac{1 + \cos 45°}{2}}\right) = 0.98.$$

Hence, $m = 0.49$ and with $l_0 \geq 0.49$, the meniscus branches will be of both signs. With $l_0 \leq 0.49$, the whole meniscus lies above the free surface level (χ_2 corresponds to the left end of the interval).

(b) The next operation, defining A from (3.19), is the most difficult one as the un-known is included in it as an argument of the elliptical functions. Let $l_0 = 1.177$, then the values of $A = 0.8$ and $\alpha_2 = \arccos 0.8 \approx 37°$ will correspond to it (Figure 3.7b, curve 6).

(c) From formulae (3.16) the coordinates of the interval ends can be found:

$$z(r_0) = -0.305, \quad z(R) = 0.736.$$

These data allow to plot a profile curve.

Formula (3.21) can be used for more accurate plotting of the curve (Figure 3.7b, curve 6).

Conclusions. Thus, based on this boundary problem solution, the following conformities can be established:

Firstly, with the angle fixation boundary condition satisfied at both ends of the r-variation interval, the vertical coordinates of the liquid-solid phase contact points with respect to the melt free surface are not fixed but depend on the relation between the angles at both ends of the interval and on the value of the clearance between the shaper and the crystal being pulled.

Secondly, transition from the wetting boundary condition on the shaper wall to the catching condition at the shaper free edge is possible in case the shaper sharp edge is placed below the melt-shaper wall contact point and its z-coordinate can be found by solving this boundary problem (Figure 3.10).

Thirdly, there exists some minimum value of the clearance between the shaper and the plate being pulled when the meniscus lies both above and below the melt free surface.

Fourthly, any change in the melt level during pulling will produce the following effect on the crystal dimensions: with the level decreasing, the tape thickness can be kept unchanged only when the crystallization front is lowered by the same value. With the crystallization front position kept unchanged, the tape thickness or the crystal diameter will decrease with the melt lowering.

3.4.2. *The Catching Boundary Conditions (Figure 3.11)*

$$z\,|_{r=R} = h, \quad h \gtrless 0 \tag{3.32}$$

$$z\,|_{r=r_0} = d, \quad d \gtrless 0, \quad h > d. \tag{3.33}$$

Solutions having both positive and negative branches, i.e., $h > 0, d < 0$, are sought. Then $z' < 0$ and the limits for the arbitrary constant A are obtained from the first integral and the boundary conditions (3.32), (3.33):

Shaped Crystal Growth

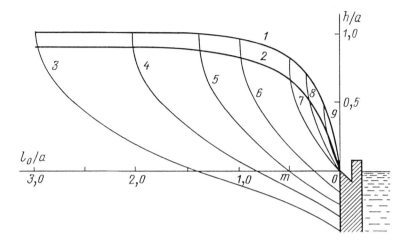

Fig. 3.10. Boundary (1, 2) and profile (3–9) curves for transition from the wetting condition (3–7) to the catching condition (7–9): $\alpha_1 = 45°$; $\alpha_0 = 90°$ (1), $\alpha_0 = 75°$ (2).

$$0 \le A \le 1 \tag{3.34}$$

Using the boundary conditions, the equation for A can be found:

$$\int_h^d \frac{A - z^2}{\sqrt{1 - (A - z^2)^2}} dz = l_0. \tag{3.35}$$

Representation of the integral (3.35) as two separate ones: from h to 0 and from 0 to d and introduction of the notations $z = \sqrt{A - \cos \alpha}$, $A = \cos \alpha_2$ gives:

$$\int_{\alpha_2}^{\arccos(h^2 - \cos \alpha_2)} \frac{\cos \alpha}{(\cos \alpha_2 - \cos \alpha)^{1/2}} d\alpha$$

$$+ \int_{\alpha_2}^{\arccos(d^2 - \cos \alpha_2)} \frac{\cos \alpha}{(\cos \alpha_2 - \cos \alpha)^{1/2}} d\alpha = 2l_0. \tag{3.36}$$

Geometrical representation of the angle α_2 follows from the previous problem: (3.18) and Figure 3.11.

As follows from (3.36), the maximum values of d^2 and h^2 are equal to *two* as there exist no solutions for $\max\{h^2, d^2\} > 2$. It means that the melt column cannot be raised above the free surface level higher than by $\sqrt{2}$. Otherwise, it breaks. The shaper cannot be immersed into the melt deeper than by $\sqrt{2}$, otherwise the melt pressure cannot be equalized by the column surface curvature and melt splashing

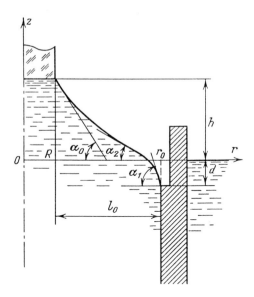

Fig. 3.11. Meniscus parameters with the catching boundary condition satisfied at shaper free edge.

from the shaper will occur. We will analyze what types of melt columns correspond to the maximum possible values of h and d.

As was noted in 1.2, in case the profile curve is described by the function $z = f(r)$, different differential equations of the (1.12)–(1.14)-types, where the sign before the last term changes depending on the sign of z', correspond to different branches of the profile curve. However, Equation (1.13) possesses an important feature. The solution of this equation is represented as the function $r = f(z)$ that will be unambiguous even in case the function $z = f(r)$ is ambiguous. Therefore, the multivalued profile curves $z = f(r)$ will be analyzed. We excluded such curves from consideration in the previous problem as either $\alpha_1 > \pi/2$ or $\alpha_0 > \pi/2$ for these curves. The first case corresponds to a horizontal- or inclined-walled nonwettable shaper when on-wall shaping is practically absent. The second case corresponds to negative growth angles. Such profile curve behavior will be explained in the present and following sections.

The maximum possible values of h and d correspond to $\alpha_2 = 0, \arccos(h^2 - \cos\alpha_2) = 0, \arccos(d^2 - \cos\alpha_2) = 0$ (Figure 3.12).

Thus, *firstly*, the melt column height described by (1.13), when measured off from the shaper level, corresponding to the maximum h and d values is equal to $2\sqrt{2}$ (curve ABCD in Figure 3.12).

Secondly, if only profile curves with positive growth angle are analyzed, the maximum melt column height is equal to $\sqrt{2} + 1$ (curve ABC in Figure 3.12).

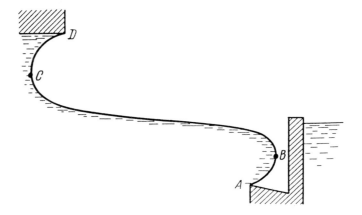

Fig. 3.12.   The profile curve of maximum height.

Thirdly, the maximum possible height of the unambiguous profile curve is equal to 2 (curve BC in Figure 3.12).

Note that these are the maximum values of h and d regardless of the parameter l_0 and they are not analyzed for stability.

The left side of (3.36) is a monotonely decreasing function of α_2. Hence, the minimum value of this expression, $l_{0\,min}$, corresponding to the maximum value of α_2 exists. Then, $l_{0\,min}$ should be found. Using (3.16), $h = \sqrt{\cos \alpha_2 - \cos \alpha_0}$, $\mid d \mid = \sqrt{\cos \alpha_2 - \cos \alpha_1}$ can be calculated. For unambiguous profile curves $z(r)$ the values of α_0 and α_1 should not exceed $\pi/2$. Let $\mid h \mid > \mid d \mid$; then as l_0 decreases, the angle α_0 will be the first to reach the value of $\pi/2$. In this case $\cos \alpha_2 = h^2$, $\cos \alpha_1 = h^2 - d^2$,

$$2l_0 \geq \int_{\arccos(h^2)}^{\pi/2} \frac{\cos \alpha}{(\cos \alpha_2 - \cos \alpha)^{1/2}} d\alpha$$

$$+ \int_{\arccos(h^2)}^{\arccos(h^2 - d^2)} \frac{\cos \alpha}{(\cos \alpha_2 - \cos \alpha)^{1/2}} d\alpha. \tag{3.37}$$

With $\mid d \mid > \mid h \mid$, h and d should be interchanged.

Hence, with h and d specified ($h > 0, d < 0$) so that they should satisfy the general inequality $\max\{h^2, d^2\} \leq 1$, the value of the clearance l_0 cannot be less than the value calculated from inequality (3.37).

Using the technique of passing to the elliptical functions, as was done for (3.18) in the previous problem, (3.36) will be obtained in the following form:

$$2G(k, \pi/2) - G(k, \varphi_d) - G(k, \varphi_h) = 2l_0. \tag{3.38}$$

Here

$$\varphi_d = \arcsin \sqrt{1 - \frac{d^2}{1 + A}}, \quad \varphi_h = \arcsin \sqrt{1 - \frac{h^2}{1 + A}}. \tag{3.39}$$

If it is assumed that $A = \max\{h^2, d^2\}$ in the left side of (3.38), then (3.38) gives the criterion (3.37) expressed in terms of the elliptical functions.

From (3.38) A can be found, and then the profile curve equation can be represented as

$$r(z) = R + 0.5[G(k, \varphi_z) - G(k, \varphi_h)], \quad z \geq 0 \tag{3.40}$$

$$r(z) = R + G(k, \pi/2) - 0.5[G(k, \varphi_h) + G(k, \varphi_z)], \quad z \leq 0. \tag{3.41}$$

The results of the present section can be used for such pulling versions when the crystallization front lies below the melt free surface level or the shaper free edge lies above the free surface level. In the former case $h < 0$, and in the latter case $d > 0$.

Conclusion. This boundary problem formulation (the catching boundary conditions) leads to the following conclusions of practical interest.

Firstly, the distance from the melt-shaper and the melt-crystal (seed) contact points, h and d respectively, to the melt free surface are related with the shaper-crystal clearance value l_0 by (3.38). The larger the value of l_0, the larger possible values of h and d. The limiting values of these quantities are calculated, this makes such problem formulation interesting.

Secondly, (3.40) and (3.41) allow plotting of the profile curve, and the interrelation of the present problem with the previous one (3.16) allows to calculate the angles (Figure 3.11) the melt makes with the shaper (α_1) and with the crystal being pulled (α_0).

3.4.3 *The catching boundary condition at the shaper free edges and angle fixation are at the crystal-melt boundary.* (Figure 3.11)

$$z\mid_{r=r} = d, \quad z'\mid_{r=R} = -tg\alpha_0. \tag{3.42}$$

Let $d < 0$ (the shaper free edge is below the melt level). The equation for the integration constant A can be written using (3.15):

$$l_0 = \int_0^{\sqrt{A - \cos\alpha}} \frac{A - z^2}{\sqrt{1 - (A - z^2)^2}} dz + \int_0^d \frac{A - z^2}{\sqrt{1 - (A - z^2)^2}} dz. \tag{3.43}$$

Substitution of $z = \sqrt{A - \cos\alpha}$ into (3.43) gives the latter in the form of (3.18) with

$$\alpha_1 = \arccos(\cos\alpha_2 - d^2). \tag{3.44}$$

Consider the way the profile curve changes as a function of l_0. If $l \to \infty$, $\alpha_2 \to 0$. In this case the angle $\alpha_{1\,min} = \arccos(1 - d^2)$ and the liquid column height $h = \sqrt{1 - \cos\alpha_0}$. The cases of $\alpha_0 > \alpha_{1\,min}$ and $\alpha_0 < \alpha_{1\,min}$ should be distinguished.

If $\alpha_0 > \alpha_{1\,min}$, with l_0 being large enough, only one value of l_0 satisfying the boundary conditions exists (Figure 3.13a). Based on the solution of the previous problem, a conclusion can be drawn that as l_0 decreases, α_2 increases, and hence, α_1 also increases. As soon as the condition $\alpha_1 > \alpha_0$ holds, two values of l_0 satisfying the boundary conditions appear: one of them (l_1) corresponds to a convex- concave meniscus, as before, and the other one (l_2) corresponds to a convex meniscus (Figure 3.13a). With this, the maximum value of $l_0 = l_0^*$ for which convex meniscus do not yet exist, corresponds to $\alpha_1 = \alpha_0$ (Figure 3.13a), $\alpha_2 = \arccos(d^2 + \cos\alpha_0)$:

$$l_0^* = \int_{\arccos(d^2 - \cos\alpha_0)}^{\alpha_0} \frac{\cos\alpha}{\sqrt{d^2 + \cos\alpha_0 - \cos\alpha}} d\alpha \tag{3.45}$$

With α_2 increasing, l_1 decreases and l_2 increases. The minimum value of l_1 equal to

$$2l_{1\,min} = \int_{\arccos(d^2)}^{\alpha_0} \frac{\cos\alpha}{\sqrt{d^2 - \cos\alpha}} d\alpha - \int_{\arccos(d^2)}^{\pi/2} \frac{\cos\alpha}{\sqrt{d^2 - \cos\alpha}} d\alpha \tag{3.46}$$

corresponds to the maximum value of l_2

$$2l_{2\,max} = \int_{\arccos(d^2)}^{\pi/2} \frac{\cos\alpha}{\sqrt{d^2 - \cos\alpha}} d\alpha \tag{3.47}$$

If $\alpha_0 < \alpha_{1\,min} = \arccos(1 - d^2)$, for sufficiently large values of l_0 two solutions satisfying the boundary conditions exist: one of them (l_1) corresponds to a convex-concave meniscus, the other one (l_2) corresponds to a convex meniscus (Figure 3.13b). As l_1 decreases, α_2 and l_2 increase. As soon as $\alpha_2 = \alpha_0$, the values of l_1 and l_2 become comparable. In this case

$$2l_1 = \int_{\alpha_0}^{\arccos(\cos\alpha_0 - d^2)} \frac{\cos\alpha}{\sqrt{\cos\alpha_0 - \cos\alpha}} d\alpha \tag{3.48}$$

The whole meniscus lies below the melt level, its upper point coinciding with the melt level.

As in the previous problems, the Legendre elliptical functions of the first and second kinds will be used. For convex-concave profile curves, (3.43) will assume the form of (3.38) if φ_h is substituted by φ_0, after that the profile curve can be plotted using (3.21).

For convex profile curves:

$$2l = G(k, \varphi_0) - G(k, \varphi_d), \tag{3.49}$$

$$r(z) = R + \frac{1}{2}[G(k, \varphi_0) - G(k, \varphi_z) \tag{3.50}$$

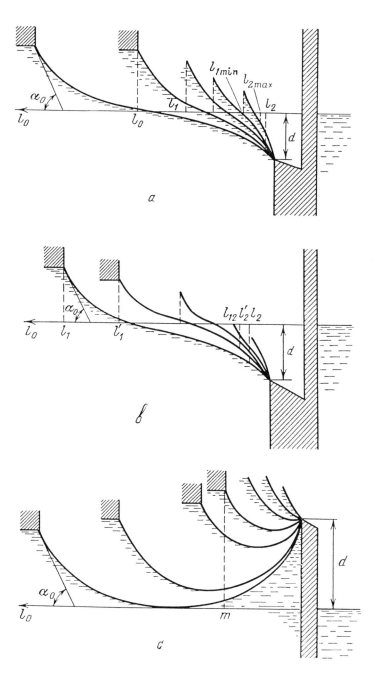

Fig. 3.13. Profile curve patterns for various values of l_0: $\alpha_0 > \arccos(1 - d^2)$ (a); $\alpha_0 < \arccos(1 - d^2)$ (b); $d > 0$ (c).

*Let $d > 0$ (the shaper free edge is above the melt level). Then for sufficiently large
values of l_0 (3.18) will assume the following form:*

$$2l_0 = \int_0^{\alpha_0} \frac{\cos \alpha}{\sqrt{A - \cos \alpha}} d\alpha + \int_0^{\alpha_1} \frac{\cos \alpha}{\sqrt{A - \cos \alpha}} d\alpha \qquad (3.51)$$

If the value of l_0 is very large, $A = \cos \alpha_1 \approx 1$ and l_0 can be represented in terms of
the elementary functions from (3.51) by analogy with (3.24) (Figure 3.13c).

As a rule, small clearances l_0 between crystals and shaper free edges are of practical
interest. As follows from (3.15), l_0 is a decreasing function of the parameter A. The
minimum value of A with $d > 0$ and $h > 0$ is equal to one. With $1 < A < 1 + d^2$,
two profile curve branches exist. With $A = 1 + d^2$, the angle α_1 is equal to *zero*,
and the right branch disappears (Figure 3.13c). The following values of l_0 and h
correspond to it:

$$2l_0 = 2m = \int_0^{\alpha} \frac{\cos \alpha}{\sqrt{1 + d^2 - \cos \alpha}} d\alpha \qquad (3.52)$$

$$h = \sqrt{1 + d^2 - \cos \alpha_0} \qquad (3.53)$$

Rearrangement of (3.52) similar to that of (3.22) gives

$$m = \frac{1 + d^2}{\sqrt{2 + d^2}} \left[F(k_d, \varphi_0^*) - F(k_d, \pi/2) \right] -$$

$$- \sqrt{2 + d^2} \left[E(k_d, \varphi_0^*) - E(k_d, \pi/2) \right] \qquad (3.54)$$

Here $k_d^2 = 2/(2 + d^2)$, $\varphi_0 = \arcsin \sqrt{1/2(1 + \cos \alpha_0)}$.

For $l_0 < 0$:

$$2l_0 = \int_{\arccos(A - d^2)}^{\alpha_0} \frac{\cos \alpha}{\sqrt{A - \cos \alpha}} d\alpha. \qquad (3.55)$$

From (3.55) l_0 is an increasing function of the parameter A. The minimum A value is
equal to $d^2 + \cos \alpha_0$, which corresponds to $l_0 = 0$, the maximum value of $A = d^2 + 1$,
which corresponds to $l_0 = m$. For this A-value interval, (3.55) can be written down in
the form of the following set of elliptical functions:

$$l_0 = \frac{A}{\sqrt{d + 1}} \left[F(k_d, \varphi_0^*) - F(k_d, \varphi_A) \right]$$

$$- \sqrt{A + 1} \left[E(k_d, \varphi_0^*) - E(k_d, \varphi_A) \right]. \qquad (3.56)$$

Here $\varphi_A = \arcsin \sqrt{1/2(1 + A - d^2)}$.

The expression for the maximum meniscus height for the case of $l_0 \rightarrow 0$ can be
found. In this, $A \rightarrow \cos \alpha + d^2$ and (3.55) assumes the following form:

$$2l_0 = d^{-1} \int_{\arccos(A-d^2)}^{\alpha_0} \cos\alpha \, d\alpha = d^{-1}[\sin\alpha_0 - \sqrt{1-(A-d^2)^2}]$$

For the integration constant the expression $A = d^2 + \sqrt{1-(\sin\alpha_0 + 2dl_0)^2}$ is obtained; hence, it is obvious that the approximation $l_0 \to 0$ can be used for $l_0 \ll 1/2d$. For the melt column height

$$h = \sqrt{d^2 - \cos\alpha_0 + \sqrt{1-(\sin\alpha_0 - 2dl_0)^2}}. \tag{3.57}$$

If the growth angle is equal to zero,

$$h \approx d + \sqrt{l_0/d - l_0^2} \tag{3.58}$$

The existence ranges for one- and two-valued meniscus can be calculated from the analysis of the first integral (3.15). The first integral can be written down in the following form: $\cos\alpha = A - z^2$. The arbitrary constant $A = \cos\alpha_0 + h^2$. Hence, $\cos\alpha_1 = \cos\alpha_0 + h^2 - d^2$. If sagging meniscus are not analyzed ($r = f(z)$ is an ambiguous function), α_1 can vary within the range of 0 to π, and the meniscus-existence range is sought from the following condition:

$$-1 \le \cos\alpha_0 + h^2 + d^2 \le 1 \tag{3.59}$$

Within the range of $0 \le \cos\alpha_0 + h^2 - d^2 \le 1$ the meniscus are two-valued. The results obtained are presented in Figure 3.14. Figure 3.15 gives a plot of meniscus for $\psi = 12°(\alpha_0 = 78°)$ under various pressures as an example.

3.5. Capillary Problem for Systems of Large Bond Numbers with an Allowance for Surface-Tension Gradients [118]

In all the above sections the surface tension at various meniscus points was considered to be constant. In practice, the meniscus always lies in a nonuniform temperature field, which leads to changes in melt surface tension. Nonuniform impurity distribution in the melt can also cause changes to its surface tension. The problem of the surface tension gradient effect on the meniscus shape ignoring convection that occurs in this case (the Marangoni convection) and pressure changes resulting from this convection were considered in our earlier work [119]. However, this very much simplified problem formulation does not reveal the main phenomena that result from variable surface tension.

Consider plate growth by TPS. Meniscus melt motion can be described by the Navier-Stokes equation with corresponding boundary conditions [63]:

$$\frac{\partial V}{\partial t} + (V\Delta)V = -\frac{1}{\rho}\,\text{grad}\,P + \frac{\eta}{\rho}\Delta V. \tag{3.60}$$

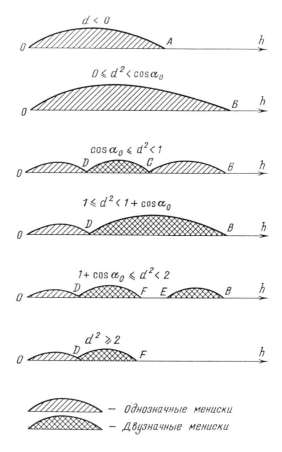

Fig. 3.14.   Existence ranges of one- and two-valued meniscus for various pressures d applied. The meniscus height h for the points A–F is as follows: A: $d + (d^2 + 1 - \cos\alpha_0)^{1/2}$; B: $d + (1 - \cos\alpha_0)^{1/2}$; C: $d + (d^2 - \cos\alpha_0)^{1/2}$; D: $d - (d^2 - \cos\alpha_0)^{1/2}$; E: $d + d^2 - 1 - \cos\alpha_0)^{1/2}$; F: $d - (d^2 - 1 - \cos\alpha_0)$.

With an allowance for surface tension gradient existence, the following boundary condition should be satisfied on the meniscus free surface:

$$\left[P_1 - P_2 - \gamma \left(\frac{1}{R_1} + \frac{1}{R_2} \right) \right] n_i = (\sigma_{ik}^{1(1)} - \sigma_{ik}^{1(2)}) n_k + \frac{\partial\gamma}{\partial x_i} \qquad (3.61)$$

where V is the flow rate, ρ denotes the liquid density, P is the pressure, η is the viscosity, $\mathbf{n}$ is the normal vector inwards the liquid, γ is the surface tension gradient, t is the time, R_1 and R_2 are the main meniscus-curvature radii and for a plate $R_2 \to \infty$, σ_{ik} is the viscous stress tensor, x_i is the coordinate. The problem will be solved with the following approximations.

Firstly, only the steady-state liquid flow will be considered, i.e., the stationary problem $\partial V / \partial t = 0$ will be solved. Secondly, by substituting the true meniscus

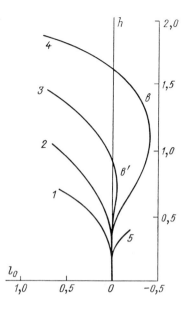

Fig. 3.15.   Boundary curves for growth angle $\psi_0 = 12°$ and different pressures d: $1 - d = 0.01$; $2 - d = 0.41$; $3 - d = 0.81$; $4 - d = 1.21$; $5 - d = 2.01$.

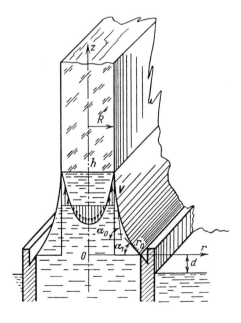

Fig. 3.16.   Parameters of real and idealized meniscus; melt-rate pattern.

shape by the shape shown in Figure 3.16 the rate and pressure distribution in the meniscus will be sought, and then using the pressure distribution found the Laplace equation will be solved and the meniscus shape will be found. In this approximation both sides of (3.61) are equal to *zero*. The left side represents the Laplace equation defining the meniscus true shape, and the right one represents the boundary condition to solve (3.60). In the coordinate system shown in Figure 3.16 Equation (3.60) has the following form:

$$\frac{\eta}{\rho}\frac{\partial^2 V}{2} = \frac{1}{\rho}\frac{\partial P}{\partial z} + g \tag{3.62}$$

where r and z are the coordinates, and g is the gravitational acceleration. Note that the term $(V\Delta)V$ is absent in this equation. In the case of nonplane flow this term is proportional to the square of the rate value and can be ignored for small rates. It is this condition, together with small film thickness, that allows the application of the approximation used.

The boundary condition of the given problem can be written down. As the problem is symmetric relative to the point where $r = 0$, it is obvious that

$$\frac{\partial V}{\partial r}\Big|_{r=0} = 0 \tag{3.63}$$

On the free surface, i.e., for $r = R$, we have:

$$\eta\frac{\partial V}{\partial r}\Big|_{r=R} = \frac{\partial \gamma}{\partial z} \tag{3.64}$$

where R is the plate half-thickness.

Consider the case when the melt free surface level coincides with the shaper level. Then on this level the pressure is equal to the ambient pressure P_0:

$$P\,|_{z=0} = P_0 \tag{3.65}$$

And finally, the melt flow $2Q_0$ through the meniscus cross-section will be specified:

$$\int\limits_{0}^{R} V(r)dr = Q_0 \tag{3.66}$$

Assume that $\gamma(z)$ varies linearly as a function of the coordinate z:

$$\gamma(z) = \gamma_0 - (\gamma_0 - \gamma_T)(h - z)/h. \tag{3.67}$$

In this case the total change of γ on the meniscus is small, i.e.,

$$\frac{\Delta\gamma}{\gamma} = \frac{\gamma_0 - \gamma_T}{\gamma_0} \ll 1. \tag{3.68}$$

Here γ_T and γ_0 denote the values of γ at the points where $z = 0$ and $z = h$, respectively (h is the meniscus height).

This assumption holds for small temperature gradients over the meniscus. The solution of the problem formulated has the following form:

$$P(z) = \left(\frac{1}{R} \frac{\partial \gamma}{\partial z} - \rho g \right) z + P_0 \tag{3.69}$$

$$V(r) = \frac{Q_0}{R} + \frac{1}{2\eta R} \frac{\partial \gamma}{\partial z} \left(r^3 - \frac{R^2}{3} \right). \tag{3.70}$$

To find the meniscus shape the Laplace equation should be solved. The problem with the gravity force absent, when the influence of surface tension inconstancy on the meniscus shape is most obvious will be analyzed. Then, for a plate in the coordinates shown in Figure 3.16 the Laplace capillary equation assumes the following form:

$$\frac{r''}{(1 + r'^2)^{3/2}} = \frac{P_0 - P(z)}{\gamma(z)}. \tag{3.71}$$

With an allowance for (3.69), it gives:

$$\frac{r''}{(1 + r'^2)^{3/2}} = \left(\rho g - \frac{1}{R} \frac{\partial \gamma}{\partial z} \right) \frac{z}{\gamma(z)} \tag{3.72}$$

Expansion of the right side of (3.72) into a series accurate up to the first-order terms in the small parameter $\Delta\gamma/\gamma_0$ gives:

$$\frac{r''}{(1 + r'^2)^{3/2}} = \frac{2z}{a^2} \left(1 - \frac{\Delta\gamma}{\gamma_0} \frac{a^2}{2Rh} + \frac{\Delta\gamma}{\gamma_0} \frac{h - z}{h} \right) \tag{3.73}$$

where $a = \sqrt{2\gamma_0/\rho g}$ is the capillary constant.

With the surface tension gradient absent, this equation for various boundary conditions was solved in the previous paragraph. The corrections to it will be sought here. Note that firstly, the second term in the parentheses of the right side of (3.73) defined by pressure changes in the system does not depend on z and in practice leads to capillary constant changes:

$$a_{ef}^2 = \left| \frac{a^2}{1 - \frac{\Delta\gamma}{\gamma_0} \frac{a^2}{2Rh}} \right|. \tag{3.74}$$

However, with an assumption that $\frac{\Delta\gamma}{\gamma} \frac{a^2}{2Rh} \ll 1$ both the second and the third terms can be regarded as small corrections. Now, the boundary condition to solve (3.73) can be written. The catching condition is believed to be satisfied at the shaper free edge (Figure 3.16).

$$r(z) \mid_{z=0} = r_0 \tag{3.75}$$

The angle fixation boundary condition is satisfied at the crystallization front (Figure 3.16):

$$r'(z) \mid_{z=h} = -\text{ctg}\alpha_0. \tag{3.76}$$

Assumption that $r(z) = x(z) + \delta r(z)$, where $x(z)$ is the solution of (3.76) with $\Delta\gamma = 0$ will give the following equation for the small correction $\delta r(z)$:

$$\delta r'' - \frac{3x'x''}{(1+x'^2)^{3/2}}\delta r' = \frac{\Delta\gamma}{\gamma_0}\frac{z}{h}\left[\frac{2(h-z)}{a^2} - \frac{1}{R}\right](1+x'^2)^{3/2} \tag{3.77}$$

with the following boundary conditions resulting from (3.75) and (3.76):

$$\delta r(z)\,|_{z=0} = 0, \quad \delta r'(z)\,|\,z = h = 0 \tag{3.78}$$

Solution of the problem formulated gives:

$$\delta r(z) = \frac{\Delta\gamma}{\gamma_0}(3ha^2 R\sin^3\alpha_{av})^{-1}$$

$$\times\left[hz^3 R - \frac{1}{2}z^4 R - \frac{1}{2}z^3 a^2 + h^3 zR + \frac{3}{2}h^2 za^2\right] \tag{3.79}$$

where α_{av} is the average value of the angle between the line tangent to the meniscus and the horizontal: $\alpha_1 < \alpha_{av} < \alpha_0$; α_1 being the angle made by the meniscus and the shaper (Figure 3.16).

Maximum deviation of r in its absolute value is reached for $z = h$:

$$\delta r(h) = \frac{1}{3}\frac{\Delta\gamma}{\gamma_0}\frac{h^2}{a^2}\frac{1}{R\sin^3\alpha_{av}}\left[a^2 - \frac{hR}{2}\right]. \tag{3.80}$$

Usually $2a^2 \gg hR$. Assumption that $\alpha_{av} = \alpha_1$ gives the upper boundary condition for δr:

$$\delta r < \frac{1}{3}\frac{\Delta\gamma}{\gamma_0}\frac{h^2}{R\sin^3\alpha_1} \tag{3.81}$$

Now the result obtained will be discussed. As follows from (3.69), presence of the surface tension gradient leads to pressure changes in the meniscus. The additional term is opposite in sign to the hydrostatic pressure and when the gravitational force is absent it plays the leading role in meniscus shaping. This pressure presence practically leads to a change in the melt capillary constant (3.74).

Two cases should be distinguished here. If $\frac{\Delta\gamma}{\gamma}\frac{a^2}{2Rh} < 1$, the effective capillary constant increases. It results in a meniscus curvature decrease. The curvature becomes equal to *zero* if $\frac{\Delta\gamma}{\gamma_0}\frac{a^2}{2Rh} = 1$, for $\frac{\Delta\gamma}{\gamma_0}\frac{a^2}{2Rh} > 1$ the curvature changes its sign. In the case of plate growing, this means changing of the concave meniscus for a convex one.

With the gravitational force absent, and $a^2 \to \infty$ as follows from (3.73) and (3.74), all the results of the previous analysis can be used to find the meniscus shape by formally substituting a^2 by $-\frac{2Rh\gamma}{\Delta\gamma}0$. However, the fact that in this case the square of the capillary constant is a negative value results in convex melt column formation when growing a plate even without additional pressure.

The third term of (3.73) describes inconstancy of the surface tension over the meniscus. Its sign is opposite to that of the second additional term and when the plate is pulled it increases meniscus concavity. Relative effect of these two terms is determined by the sign of (3.80). For $a^2 > 1/2Rh$, the effect of the second term prevails, while for $a^2 < 1/2Rh$, the effect of the third term does.

As follows from (3.70), against the background of the total liquid flow Q_0/R possesses convective liquid flow for the rate distribution shown in Figure 3.16. This flow is the result of the surface tension gradient and its contribution to the total flow decreases with the surface tension gradient decreasing.

In the previous discussion of the results obtained it was assumed that $\Delta\gamma > 0$, i.e., $\gamma_0 > \gamma_T$, which corresponds to overheated melt. In the case of supercooled melt, all the formula forms remain unchanged, but $\Delta\gamma < 0$. It corresponds to the coincidence of the sign of the pressure increment and that of the hydrostatic pressure, and in this case the capillary constant decreases, i.e., in the case of plate growing meniscus concavity increases. The third term of (3.73) with $\Delta\gamma < 0$ increases meniscus convexity. The convective flow pattern also changes. Now the liquid flows "down" at the meniscus edge and "up" in its center.

3.6. The Capillary Problem for Systems of Small Bond Numbers

Equation (1.14) describing the meniscus shape in the process of growing crystals with the diameter smaller than the capillary constant (a small Bond number) [31, 120–122] will be analyzed. The limits of this equation applicability will be discussed below.

3.6.1. *The Pressure Is Equal to Zero*

Such problem formulation is possible only for the catching boundary condition satisfied at the shaper free edge. The angle fixation condition at the crystal-melt boundary is specified as the second boundary condition. Therefore, the following boundary problem can be formulated:

$$z''r + z'(1 + z'^2) = 0 \tag{3.82}$$

$$z\,|_{r=r} = 0 \tag{3.83}$$

$$z\,|_{r=R} = -tg\alpha_0 \tag{3.84}$$

Note that if the crystal contour projection lies inside the shaper ($R < r_0$), the first derivative is negative and vice versa (Figure 3.17). The second derivative in the profile curve equation (3.82) is opposite in sign to the first one.

We should define more accurately what is meant by concavity and convexity of profile curves. A profile curve is regarded as concave if the center of its curvature lies outside the liquid and convex if the center of its curvature lies inside the liquid. Therefore, regardless of the sign of the second derivative profile curves will be concave in both the cases ($R < r_0$ and $R > r_0$). In the latter case growth angles have to be negative.

$R < r_0$. The first integral of (3.82) has the following form:

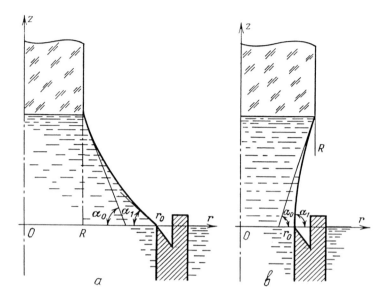

Fig. 3.17. Profile curve shapes for $R < r_0$ (a); for $R > r_0$ (b).

$$\frac{r z'}{\sqrt{1 + z'^2}} = C_1. \tag{3.85}$$

As follows from (3.84), $C_1 = R \sin \alpha_0$. The profile curve equation is obtained in the following form:

$$z(r) = R \sin \alpha_0 \left(\text{arch} \frac{r_0}{R \sin \alpha_0} - \text{arch} \frac{r}{R \sin \alpha_0} \right). \tag{3.86}$$

Using the angle α_1 (Figure 3.17) an expression for the profile curve height can be obtained:

$$z(R) = r_0 \sin \alpha_1 \left(\text{arch} \frac{1}{\sin \alpha_1} - \text{arch} \frac{1}{\sin \alpha_0} \right). \tag{3.87}$$

With α_1 fixed, the right side of (3.78) is a monotone α_0-function. Unambiguous profile curves will be analyzed. In this case α_0 varies within the limits of α_1 to $\pi/2$. With $\alpha =_0 \pi/2$, the profile curve will be called **limiting** (Figure 3.18) and its finite point abscissa will be denoted as R_0. Then this profile curve equation will be as follows:

$$z(r) = R_0 \left(\text{arch} \frac{r_0}{R} - \text{arch} \frac{r}{R_0} \right) \tag{3.88}$$

and its height can be expressed in the following way:

$$h_0 \equiv z(R_0) = R_0 \, \text{arch} \frac{r_0}{R_0} \tag{3.89}$$

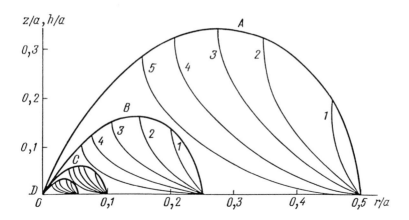

Fig. 3.18. Limiting profile curves $z(r) - 1, 2, 3, 4, 5$ and boundary curves $h(R) \mid_{\alpha=90} 0 - A, B, C, D$ for $r_0 = 0.5$ (A), $r_0 = 0.25$ (B), $r_0 = 0.1$ (C), $r_0 = 0.05$ (D); the pressure is equal to zero.

Figure 3.18 shows a family of limiting profile curves for various shaper hole-to-crystal dimension ratios (curves 1–5). Therein the **boundary curves** representing the loci of the ends of the profile curves characterized by the same α_0-values, i.e., providing the growth angle value $\psi_0 = \pi/2 - \alpha_0$ are plotted. The boundary curve equation is as follows:

$$h(R) \equiv z(R) \mid_{\alpha_0=\text{constant}} = R \sin \alpha_0 \left(\text{arch} \frac{r_0}{R \sin \alpha_0} - \text{arch} \frac{1}{\sin \alpha_0} \right) \quad (3.90)$$

For the limiting curves this equation goes over into (3.89). The area inside the boundary curve is the range of column existence with the growth angle $\psi_0 \geq \pi/2 - \alpha_0$ for a given shaper under the condition of melt column catching on the sharp shaper free edge. The boundary curve exhibits a peak value. A profile curve of the largest height z_m for a given shaper dimension r_0 corresponds to the peak-value point R_m. The peak-value position of the boundary curve (R_m, z_m) of (3.89) can be calculated from the following equation:

$$\frac{dz}{dR_0} \mid_{R_0=R_m} = \text{arch} \frac{r_0}{R_m} - \frac{R_m^2}{\sqrt{r_0^2 - R_m^2}} = 0 \quad (3.91)$$

The numerical solution of this equation is given in [68]:

$$R_m = 0.55 r_0, \quad z_m = 0.66 r_0 \quad (3.92)$$

From the expression of the first integral (3.85), the value of the angle made by the profile curve with the horizontal at the shaper free edge can be found:

$$\alpha_1 = \arcsin \left(\frac{R}{r_0} \sin \alpha \right) \quad (3.93)$$

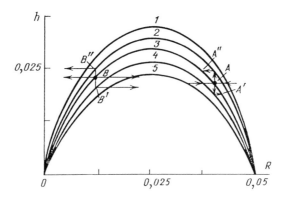

Fig. 3.19. Boundary curves $h(R)|_{\alpha=const}$ for $r_0 = 0.05$; $1 - \alpha_0 = 90°$; $2 - \alpha_0 = 85°$; $3 - \alpha_0 = 80°$, $4 - \alpha_0 = 75°$; $5 - \alpha_0 = 70°$; A – the point of capillary stability, B – the point of capillary instability.

For the limiting profile curve:

$$\alpha_1 = \arcsin(R_0/r_0) \tag{3.94}$$

Figure 3.19 shows a family of the limiting curves $h|_{\alpha_0} = z(R)|_{\alpha_0=const}$, plotted from (3.90). The behavior of these curves will be necessary to study crystallization process stability.

$R > r_0$. The solution of (3.82) for $z' > 0$ is sought (Figure 3.17). For the first integral the following expression is derived:

$$rz'(1 + z'^2)^{-1/2} = C_1 \tag{3.95}$$

The profile curve equation will take the following form:

$$z(r) = R \sin \alpha_0 \left(\text{arch} \frac{r}{R \sin \alpha_0} - \text{arch} \frac{r_0}{R \sin \alpha_0} \right). \tag{3.96}$$

The boundary curve equation will be as follows:

$$h(R) = R \sin \alpha_0 \left(\text{arch} \frac{1}{\sin \alpha_0} - \text{arch} \frac{r_0}{R \sin \alpha_0} \right). \tag{3.97}$$

Equation for $h(R)$ can be represented in a slightly changed form:

$$h(R) = r_0 \sin \alpha_1 \left(\text{arch} \frac{r}{R \sin \alpha_0} - \text{arch} \frac{1}{\sin \alpha_1} \right). \tag{3.98}$$

As follows from (3.98), $h(R)$ is a monotonic function α_1. For $\alpha_1 = \pi/2$ this profile curve will be called an *extreme* one. The equation of this curve is:

$$z(r) = r_0 \, \text{arch} \frac{r}{R \sin \alpha_0} \tag{3.99}$$

3.6.2. *The Pressure Is Non-Zero* $R < r_0$

Equation (1.14) is analyzed using the catching boundary condition at the shaper free edge and the angle fixation condition at the melt-crystal boundary. The z-coordinate origin lies at the shaper free edge:

$$z \mid_{r=r_0} = 0, \tag{3.100}$$

$$z' \mid_{r=R} = -tg\alpha_0 \tag{3.101}$$

The existence ranges for profile curves of various forms. Equation (1.14) will be represented in the following form:

$$z'' = -\frac{1}{r}(1 + z'^2)[z' + 2dr(1 + z'^2)^{1/2}] \tag{3.102}$$

For crystals of smaller diameters than that of the shaper, $z' < 0$. Therefore, if the pressure is negative or equal to zero ($d \leq 0$), $z'' > 0$ in the whole range of r-values and the curve is concave (Figure 3.20 a,d). If the pressure is positive, the sign of the second derivative of the profile curve at the point corresponding to the shaper free edges is to be analyzed to define the profile curve type. In this case, if $z'' \mid_{r=r_0} < 0$ the profile curve is either convex (Figure 3.20c) or convex-concave (Figure 3.20b). To define the existence ranges for all the profile curve types enumerated, the first integral of (1.14) should be written down:

$$r^2 d - \frac{rz'}{\sqrt{1 + z'^2}} = r^2 d - r \sin \alpha = C_1. \tag{3.103}$$

From (3.102) and (3.103) it follows that:

$$\sin \alpha_1 = \frac{R}{r_0} \sin \alpha_0 + r_0 d \left[1 - \left(\frac{R}{r_0} \right)^2 \right] \tag{3.104}$$

Here

$$\alpha(r) = -\text{arctg}(dz/dr), \quad \alpha_1 = \alpha(r_0), \quad \alpha_0 = \alpha(R).$$

Equation (3.104) correlates four profile curve parameters: α_1, $\alpha_0 E$, R/r_0 and $r_0 d$. In the coordinate system $r_0 d$, $\sin \alpha_1$, R/r_0, the allowed-value range of these three parameters will represent a 3D-body and for some specific α_0-value, a surface crossing this body can be obtained. A particular profile curve will be represented by a point within the space of these three parameters that belongs to the surface characterized by the specified α_0-value. A specific profile-curve type will correspond to each point in the space mentioned.

The parameter ranges corresponding to various profile-curve types will be analyzed and represented in (Figure 3.21).

Ranges of the parameters will be defined.

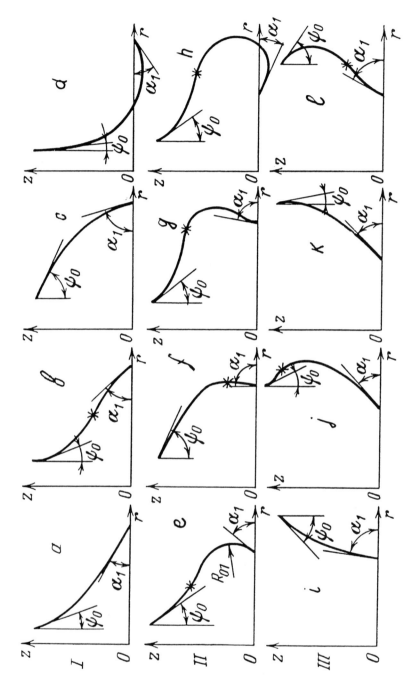

Fig. 3.20.   Various types of profile curves: 1 – one valued ones ($R < r_0$); 2 – two-valued ones ($R < r_0$); 3 – one- and two-valued ones ($R > r_0$).

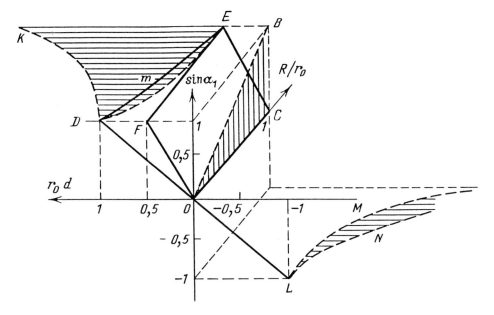

Fig. 3.21. Existence ranges of profile curves of various types in the space of the parameters R/r_0, $r_0 d$, $\sin \alpha_1$ with $0 < R/r_0 \le 1$.

As follows from the condition $R < r_0$, $R/r_0 \le 1$ and $-\pi/2 \le \alpha_1 \le \pi/2$, i.e., $-1 \le \sin \alpha_1 \le 1$. No constraints are imposed on the shaper dimensions and the pressure $-\infty \le r_0 d \le +\infty$. In this case one should remember that application of (1.14) assumes infinitesimal characteristic dimensions of the system (r_0 in this particular case) as compared with the capillary constant. Now positive growth angles $0 \le \psi_0 \le \pi/2$ are to be used in the analysis, therefore, for α_0 we obtain that: $0 \le \alpha_0 \le \pi/2$

Only concave curves proved to exist for $r_0 d < 0$ (Figure 3.21). Now it should be defined whether concave profile curves can exist under positive pressures. For convenience (3.102) will be represented in the following form:

$$z'' = -1/2(1 + z'^2)^{3/2}(-\sin \alpha + 2dr)$$ (3.105)

As follows from this equation, $z'' \mid_{r=r_0} > 0$, i.e., the profile curve is concave at the point corresponding to the shaper free edge if

$$\alpha_1 \ge \arcsin(2dr_0)$$ (3.106)

This profile curve remains concave up to the crystal free edge. It does not turn into a concave-convex one since the condition of inflection point existence $2dr - \sin \alpha = 0$ is not satisfied for (3.106) in the range of $[R, r_0]$ as $\mid \alpha \mid > \mid \alpha_1 \mid$ and $r < r_0$. Thus, for positive pressures applied irrespective of the R/r_0 value, the condition of concave profile curve existence is as follows:

$$\sin \alpha_1 \geq 2r_0 d \geq 0 \tag{3.107}$$

In Figure 3.21 this boundary position is specified by the plane OFEC. By using (3.105) the equation characterizing correlation of the parameters providing concave-curve existence for positive pressure can be obtained:

$$0 \leq d \leq R \sin \alpha_0 / (R^2 + r_0^2) \tag{3.108}$$

Now other limits of the concave curve existence range can be defined.

a) The pressure is negative but the profile curve does not sag below the shaper free edge, i.e., $\sin \alpha_1 \geq 0$ (Figure 3.20a).

$$d \leq 0, \quad 0 \leq |d| \leq R \sin \alpha_0 / (r_0^2 - R^2). \tag{3.109}$$

b) The pressure is negative, the profile curve sags below the shaper free edge (Figure 3.20d). In this, the maximum absolute value of $|\sin \alpha_1| = 1$, which corresponds to $\sin \alpha_1 = -1$. Such profile curves exist in the following range of pressure values:

$$d < 0, \quad \frac{R \sin \alpha}{r_0^2 R^2} 0 \leq |d| \leq \frac{r_0 + R \sin \alpha_0}{r_0^2 R^2} \tag{3.110}$$

With this, the profile curve first goes down, reaches its minimum at the point r_1 (Figure 3.20d):

$$r_1 = \sqrt{R^2 - \frac{R \sin \alpha_0}{d}} \tag{3.111}$$

and then rises making the angle α_0 with the horizontal at the point R.

Thus, *the limits of the concave profile curve existence range* are as follows (Figure 3.21):

In the plane $R/r_0 = 0$: $\sin \alpha_1 = r_0 d$ if $r_0 d < 0$ (straight line OL).
In the plane $R/r_0 = 0$: $\sin \alpha_1 = 2r_0 d$ if $r_0 d \geq 0$ (straight line OF).
In the plane $r_0 d = 0$: $\sin \alpha_1 = (R/r_0) \sin \alpha_0$;
In the plane $\sin \alpha_1 = 0$: $r_0 d = (-R/r_0) \sin \alpha_0 [1 - (R/r_0)^2]^{-1}$;
In the plane $\sin \alpha_1 = -1$: $r_0 d = -[1 + (R/r_0) \sin \alpha_0][1 - (R/r_0)^2]^{-1}$;
In the plane $\sin \alpha_1 = 1$: $r_0 d = 1/2$ (straight line FE).

Now the existence range of convex-concave profile curves will be defined. Convex-concave profile curves exist in such cases if one of the profile curve points is a point of inflection.

Changes in the meniscus shape with the pressure d changing continuously will be analyzed and it will be shown that the initial profile curve point of inflection appears at the shaper free edge ($r^* = r_0$). For this purpose Equation (3.105) will be used. As follows from this equation, the inflection point lies at the shaper free edge if

$$\sin \alpha_1 = 2r_0 d \tag{3.112}$$

As follows from the same equation, the inflection point lies at the crystal free edge if

$$\sin \alpha_0 = 2Rd \qquad (3.113)$$

From here the following equation can be obtained for the case under consideration with an allowance for (3.104):

$$\sin \alpha_1 = 2r_0 d[1 + (R/r_0)^2] \qquad (3.114)$$

Higher pressure than that in (3.112) satisfies (3.114). Thus, α_1 and R/r_0-values being fixed, with the pressure increasing, the inflection point first appears at the shaper free edges, then it displaces along the profile curve, and at the moment it reaches the three-phase line the whole meniscus becomes convex.

From (3.112) and (3.114) *the limits of the convex-concave profile curve existence range* can easily be found (Figure 3.21).

In the plane $R/r_0 = 0$: $r_0 d < \sin \alpha_1 < 2r_0 d$ (area DOF).

In the plane $R/r_0 = 1$: $\sin \alpha_1 = 2r_0 d$ (straight line E).

In the plane $\sin \alpha_1 = 1$: $1/2 < r_0 d < [1 - (R/r)^2]^{-1}$ (area FED).

Thus, in the space of the parameters $r_0 d$, R/r_0, $\sin \alpha_1$ the convex-concave profile curves existence range is separated from the concave curve existence range by the plane OFEC ($\sin \alpha_1 = 2r_0 d$) and by surface ODE $\{\sin \alpha_1 = r_0 d[1 + (R/r_0)^2]\}$ from the convex curve existence range. The following pressure ranges correspond to the convex-concave profile curve existence range:

$$\frac{R \sin \alpha_0}{r_0^2 R^2} \leq d \leq \frac{\sin \alpha_0}{2R} \qquad (3.115)$$

The ordinate of the inflection point r^* is given by the following expression:

$$r^* = [(\sin \alpha_0 - Rd)R/d]^{1/2} \qquad (3.116)$$

From the condition $\sin \alpha_1 \leq 1$ the limiting pressure can be found if meniscus with one-valued projections on the abscissa axis are formed:

$$d \, leq (r_0 - R \sin \alpha_0)/(r_0^2 - R^2) \qquad (3.117)$$

The growth angle constancy condition imposes additional restrictions on the allowed range of $r_0 d$-parameter values defined above. Indeed, by assuming that $\sin \alpha_1 = \mp 1$ in (3.104) it can be found that the parameter $r_0 d$ can vary in accordance with the following relation:

$$r_0 d = \frac{\mp 1 - (R/r_0) \sin \alpha_0}{1 - (R/r_0)^2} \qquad (3.118)$$

In this equation all the possible values of the angle $\alpha_0 = \pi/2 - \psi_0$ lie within the interval $0 \leq \alpha_0 \leq \pi/2$, since only positive growth angles are analyzed here. The limits of allowed $r_0 d$-value ranges in the limiting planes $\sin \alpha_1 = 1$ and $\sin \alpha_1 = -1$ ar shown in Figure 3.21 by the dashed lines, AD and DE, OH and AN, are calculated

Shaped Crystal Growth

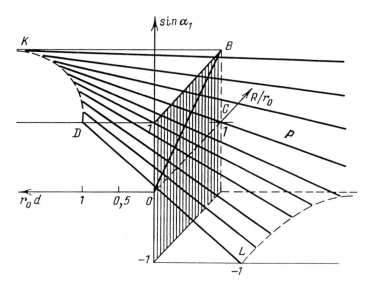

Fig. 3.22.   Surface P, a locus of the profile curves with a specified growth angle $0 < \alpha_0 < \pi/2$ in the space of parameters R/r_0, $r_0 d$, $\sin \alpha_1$.

from (3.118) for $\psi_0 = 0$, i.e., $\sin \alpha_0 = 1$, and for $\psi_0 = \pi/2$, i.e. $\sin \alpha_0 = 0$, respectively.

The meniscus existence conditions during stationary crystallization will be analyzed for one specific growth-angle value. In this case the allowed values of the parameters $\sin \alpha_1$, $r_0 d$ and R/r_0 lie on the complicated-shape surface P shown in Figure 3.22. It is obvious that (3.104) correlating the capillary parameters with α_0 and R/r_0 being fixed represents an equation of a straight line. Strictly speaking, the whole surface P (Figure 3.22) can be represented as being composed of straight lines (3.104) corresponding to various values of the parameter R/r_0 within the interval $0 < R/r_0 < 1$.

For $R/r_0 = 1$, from (3.104) it follows that $\sin \alpha_1 = \sin \alpha_0$. A stationary growing crystal of the radius $R = r_0$ can be obtained for the only pressure $d = \sin \alpha_0 / 2 r_0$. In particular, in the case of zero growth angle, $\psi_0 = 0$, the meniscus has the shape of a straight circular cylinder.

To make it more demonstrative, Figure 3.23 shows a section of the space diagram of Figure 3.21 by the plane $R/r_0 = 0.5$ for various growth angles ψ_0. The straight lines AC and BD correspond to Equation (3.104) for $\psi_0 = 0°$ and $\psi_0 = 90°$, respectively. The arrow shows the direction of point displacement along the straight line which corresponds to changes in the meniscus state introduced by an increase in pressure. As is obvious from this figure, with $r_0 d$ increasing the concave meniscus turns convex-concave, if the condition $\sin \alpha_1 = 2 r_0 d$ is satisfied and after that it can turn convex. Cases when the meniscus cannot be convex, e.g., for $\psi_0 = 0$ and $R/r_0 = 0.5$, are also possible (Figure 3.23).

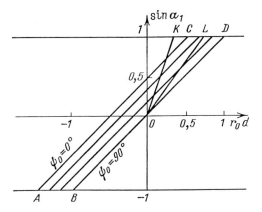

Fig. 3.23. Section of the space diagram (Figure 3.21) by the plane $R/r_0 = 0.5$.

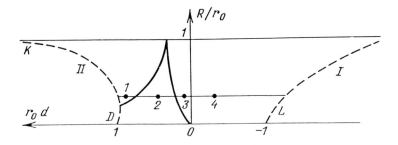

Fig. 3.24. Projection of the surface P (Figure 3.22) on the horizontal plane; 1, 2 – ranges of forbidden values.

Instead of the space diagram of Figure 3.21, with the growth angle fixed, its projection onto the horizontal plane can also be analyzed. To illustrate this, such projection for $\psi_0 = 60^0$ is plotted in Figure 3.24. The curves a and b are the lines of intersection of the surface P (Figure 3.22) described by (3.104) and the surfaces OFEC and $OD_m EC$ defining the limits of the meniscus-type change and described by the relations $\sin \alpha_1 = 2r_0 d$ and $\sin \alpha_1 = r_0 d[1 + (R/r_0)^2]$, respectively. Thus, these lines of intersection limit the existence ranges of convex (point 1), convex-concave (point 2) and concave (points 3,4) meniscus. The curves c and e are the lines of intersection of the surface P and the surfaces $\sin \alpha_1 = 1$ and $\sin \alpha_1 = -1$, respectively, i.e., I and II are forbidden ranges.

Profile-curve heights and shapes. To define the profile-curve shapes and heights the boundary problem for Equation (1.14) with the boundary conditions (3.100), (3.101) needs to be solved. Using the first-integral expression (3.103), the $z(r)$ profile curve equation is obtained in the following form:

$$z(r) = -\int_r^{r_0} \frac{C_1 - r^2 d}{\sqrt{r^2 - (C_1 - r^2 d)^2}} dr. \tag{3.119}$$

Substitution of the integration limit r by R in (3.119), gives the profile-curve height. Now (3.119) will be rearranged as follows:

$$z(r) = -\frac{1}{d}\int_r^{r_0} \frac{C_1 - r^2 d}{\sqrt{-(r^2 - \Lambda^2)(r^2 - M^2)}} dr. \tag{3.120}$$

Here

$$\Lambda^2 = \frac{1 + 2C_1 d - \sqrt{1 + 4C_1 d}}{2d^2}, \quad M^2 = \frac{1 + 2C_1 d + \sqrt{1 + 4C_1 d}}{2d^2} \tag{3.121}$$

Concave and convex-concave profile curves. Limiting $(\alpha_0 = \pi/2)$ concave and convex-concave profile curves for positive pressures applied will be analyzed. From the first integral:

$$C_1 = R_0^2 d - R_0$$

hence,

$$\Lambda^2 = R_0^2, \quad M^2 = (1/d - R_0)^2$$

A solution (the radicand in the integral (3.120) is a positive function) of our problem is possible if $\Lambda \le r \le M$. This condition is equivalent to the following problem-parameter correlation:

$$R_0 < r, r \le (1/d) - R_0.$$

For unambiguous profile curves analyzed in the given section $R_0 \le r \le r_0$, the former inequality is automatically satisfied and from the latter one the pressure range where limiting concave and convex-concave profile curves could exist can be defined: $d \le (r_0 + R_0)^{-1}$, which fits (3.117). Now the limiting profile curve shape and height can be found. For this purpose the following substitution:

$$r^2 = M^2 - (M^2 - \Lambda^2)x^2 \tag{3.122}$$

can be made and the expression for the profile curve shape (3.119) and height (Figure 3.25) in the form of a set of the Legandre elliptical functions of the first $F(k, \varphi)$ and second $E(k, \varphi)$ kinds will be obtained:

$$z(r) = R_0[F(k_0, \varphi_r) - F(k_0, \varphi_{r0})]$$

$$+\frac{1 - R_0 d}{d}[E(k_0, \varphi_r) - E(k_0, \varphi_{r0})] \tag{3.123}$$

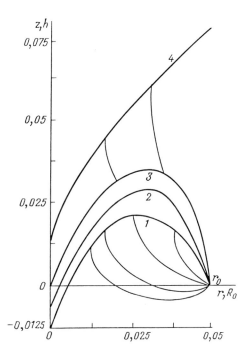

Fig. 3.25. Boundary curves $h(R_0)$ 1–4 and some profile curves $z(r)$ for shaper of $r_0 = 0.05$ under various pressures d: $1 – (-10)$; $2 – (-5)$; $3 – 0$; $4 – 10$.

$$z(R_0) = R_0 \left[F\left(k_0, \frac{\pi}{2}\right) - F(k_0, \varphi_{r0}) \right]$$

$$+ \frac{1 - R_0 d}{d} \left(E\left(k_0, \frac{\pi}{2}\right) - E(k_0, \varphi_{r0}) \right). \tag{3.124}$$

Here the Legandre modulus k_0 and the amplitudes φ_r and φ_{r0} have the following form:

$$k_0^2 = \frac{M^2 - \Lambda^2}{M^2} = \frac{1 - 2R_0 d}{(1 - R_0 d)^2}$$

$$\varphi_r = \arcsin \sqrt{1 - \frac{d^2(r^2 - R_0^2)}{1 - 2R d_0}}, \quad \varphi_{r0} = \arcsin \sqrt{1 - \frac{d^2(r_0^2 - R_0^2)}{1 - 2R_0 d}}.$$

Some particular cases will be analyzed now.

Let $R_0 = r_0$. In this case for all pressure values the profile-curve height is equal to *zero* as $\varphi_{r0} = \pi/2$ and both the expressions between the brackets in (3.124) go to *zero*. Only for $r_0 d = 0.5$, an indeterminacy in the form of 0/0 appears under the radical in the expression for φ_{r0} and by evaluating it according to the Lopital rule $\varphi_0 = \pi/4$ is obtained. In this case, the following expression is obtained for the profile curve height (Figure 3.25, curve 4):

$$h(r_0) = 0.5\pi r_0. \tag{3.125}$$

In Figure 3.21 point E corresponds to this profile-curve type.

Rod is small, R_0 *is close to* r_0, the case which often occurs in practice. Then φ_{r0} is close to $\pi/2$ and when expanding the elliptical functions into a power series [117] it can be restricted to a small number of terms and then $z(R_0)$ is obtained in the form of a set of elementary functions:

$$z(R) = \left(\frac{\pi}{2} - \varphi_{r0}\right)\left[R_0\left(1 + \frac{k_0^2}{4}\right) + \frac{1 + R_0}{d}\left(1 - \frac{k_0^2}{4}\right)\right]. \tag{3.126}$$

Nonlimiting profile curves. For nonlimiting profile curves ($\alpha_0 < \pi/2$) the expressions for k_0, Λ, M, $z(r)$ and $z(R)$ are so awkward that there is no sense in using them. It is much easier to apply the following procedure: to regard this nonlimiting profile curve as a portion of the limiting profile curve. Indeed, let the crystal diameter and the growth angle be specified, i.e., R and $\alpha_0 = \pi/2 - \psi_0$ are known. In case the crystal is pulled from a shaper with the diameter of $2r_0$ under the pressure d, with R_0 used to denote the abscissa of the limiting profile curve, from the expression for the first integral (3.103) the following equation defining R_0 can be obtained:

$$R^2 d - R\sin\alpha_0 = R_0^2 d - R_0. \tag{3.127}$$

Now the nonlimiting profile curve is a portion of the limiting profile curve limited by the abscissa $r = R$ and can easily be plotted from formula (3.123).

Negative pressure. Let $\mid d \mid = d_1$. In this instance, for limiting concave curves (it should be stressed that only concave curves can exist in this case)

$$C_1 = -R_0^2 d_1 - R_0,\ \Lambda^2 = R_0^2,\ M = (R + 1/d_1)^2.$$

From the condition of problem solution existence the limiting negative-pressure value $d_1 = 1/(r_0 - R_0)$ can be found, which agrees with the value found above (3.110).

In this case the limiting curve can be plotted from formula (3.123) with d substituted by $-d_1$ (Figure 3.25).

Convex profile curves. From (3.103) the first integral for the limiting convex profile curve can be obtained:

$$C_1 = R^2 d.$$

The solution-existence conditions are as follows: $\Lambda \le r \le M$. The first in equality gives the evident relation $\sin\alpha_0 < 1$, while the pressure ranges of melt feeding into the shaper follow from the second inequality: $d \le r_0/(r_0^2 - R^2)$, which fits the condition (3.118).

Substitution of (3.122) gives the expression for the shape $z(r)$ and the height $z(R_0)$ of the limiting profile curves in the form of a set of the Legandre elliptical functions F and E:

$$z(r) = z(k_3, \varphi_8, \varphi_9) = \frac{\sqrt{2}dR_0^2}{\sqrt{B}}[F(k_3, \varphi_8) - F(k_3, \varphi_9)]$$

$$-\frac{\sqrt{B}}{\sqrt{2d}}[E(k_3, \varphi_8) - E(k_3, \varphi_9)]. \qquad (3.128)$$

Here,

$$B = 1 + 2d^2 + R_0^2 + \sqrt{1 + 4d^2 R_0^2}$$

$$k_3^2 = \frac{2\sqrt{1 + 4d^2 R_0^2}}{1 + 2d^2 + R_0^2 + \sqrt{1 + 4d^2 R_0^2}}$$

$$\varphi_8 = \varphi(r) = \arcsin\sqrt{\frac{1}{2} + \frac{1 - 2d^2(r^2 - R_0^2)}{2\sqrt{1 + 4d^2 R_0^2}}}, \quad \varphi_7 = \varphi_8(R_0),$$

$$\varphi_9 = \varphi_8(r_0).$$

The nonlimiting profile curve can reasonably be represented as a portion of the limiting profile curve. The abscissa R_0 of the end of the corresponding limiting profile curve can be found from the first integral $R^2 dR \sin \alpha_0 = R_0^2 d$. Then the nonlimiting profile curve can be plotted using the formula (3.128).

A particular case of $R_0 \to 0$ will be considered now. If the crystal diameter is negligible, $C_1 = 0$ for both convex and convex-concave profile curves. In this case the heights of the profile curves tend to the limit specified by the following expression:

$$h = z(R_0) = \frac{1}{2d} \arcsin(r_0 d)^2. \qquad (3.129)$$

From this formula, the height of a small drop squeezed out from the shaper can be calculated in particular.

3.6.3. *The Pressure Differs from Zero $R > r_0$*

The existence ranges of profile curves of various types. Equation (1.14) will be analyzed by representing it in the form of (3.102). The solutions for which $z' > 0$ are sought. Then the minus sign should be selected before d in Equation (1.14)

$$z'' = -\frac{1}{r}(1 + z'^2)[z' - 2dr(1 + z'^2)^{1/2}] \qquad (3.130)$$

For $R > r_0$, the profile curve is concave if $z'' < 0$ (Figure 3.20i) and the profile curve is convex if $z'' > 0$ (Figure 3.20k). If the pressure is negative, $z'' < 0$ within the whole range of r-values and the profile curve is concave (Figure 3.27). If the pressure is positive, convexity or concavity of the profile curve can be estimated by the sign of the second derivative of the profile curve at the shaper free edge, as was mentioned above. Let (3.130) be represented in the following form:

$$z'' = -\frac{1}{r}(1 + z'^2)^{3/2}(\sin\alpha - 2rd).$$ (3.131)

If $\sin\alpha_1 > 2r_0d$, the profile curve at the shaper free edge is concave. For a concave curve $\sin\alpha$ should decrease with r increasing, therefore, with the pressure being positive and not equal to zero, the point where $z'' = 0$, i.e. the point of inflection, can be reached and hence the profile curve in this range can be either concave or convex-concave (Figure 3.20i, 1).

The first integral (3.103) of Equation (1.14) for concave-convex curves is as follows:

$$r_0\sin\alpha_1 - r_0^2d = r_n\sin\alpha_n - r_n^2d = R\sin\alpha_0 - R^2d,$$ (3.132)

where R is the abscissa of the profile curve peak, r_n is the abscissa of the inflection point, α_n is the angle between the abscissa axis and the line tangent to the profile curve at the point of inflection. From (3.132) the abscissa of the inflection point, r_n, can be found. From (3.131) for the inflection point we have $\sin\alpha_n = 2r_nd$, from where

$$(r_n/r_0)^2 = (\sin\alpha_1 - r_0d)/r_0d,$$

and since $\sin\alpha_1 > 2r_0d$, the inflection point can exist. Figures 3.26 and 3.27 show the ranges of concave-convex curve existence. From Figure 3.27 the value of $(r_n/r_0)^2$ for the profile curve corresponding to the point n can be calculated as the tangent of the angle in the triangle, one side of which is equal to r_0d, the other one being equal to $\sin\alpha_1 - r_0d$.

If $\sin\alpha_1 < 2r_0d$, the profile curve at the shaper free edge is convex (Figure 3.20j, k). For a convex profile curve $\sin\alpha$ should increase with R increasing. If there exists a situation when $\sin\alpha_n = 2r_nd$, this point of the profile curve will be the point of inflection. Just as in the previous case, for the point of inflection

$$(r_n/r_0)^2 = (\sin\alpha_1 - r_0d)/r_0d,$$

but since $\sin\alpha_1 < 2r_0d_n$, then $r_n < r_0$ but this does not agree with the conditions of our problem. Hence, convex-concave profile curves (Figure 3.20j) cannot exist and, for $\sin\alpha_1 < 2r_0d$, all the profile curves are convex. This statement is also valid for $\alpha_1 < 0$. In this case, the profile curve first goes down and then goes up (Figure 3.20h). The abscissa R_0^* of its minimum can be found from (3.132):

$$(R_0^*/r_0)^2 = (r_0d - \sin\alpha_1)/r_0d.$$

Pressure-limit variations for profile curves of various types. The first integral (3.104) can be analyzed to calculate them.

For convex profile curves with $\sin\alpha_1 < 0$ (Figure 3.20h) the maximum pressure corresponds to

$$r_0d = \frac{(R/r_0)\sin\alpha_0 + 1}{(R/r_0)^2 - 1}.$$ (3.133)

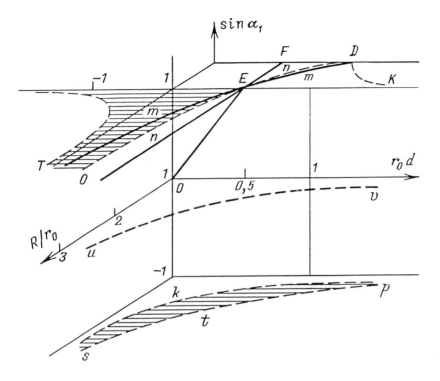

Fig. 3.26. Existence ranges for profile curves of various types within the parameters R/r_0, r_0d, $\sin\alpha_1$ with $R/r_0 > 0$. The diagram of Figure 3.21 forms part of $R/r_0 < 1$.

Expression (3.133) defines the range of profile curve existence in the plane $\sin\alpha_1 = -1$ (Figure 3.26). The limits of this range correspond to the limiting values of α_0, i.e., *StP* (Figure 3.26), where

$$r_0d = \frac{1}{(R/r_0 - 1)}, \quad \text{if } \alpha_0 = \pi/2. \tag{3.134}$$

SkP (Figure 3.26), where

$$r_0d = \frac{1}{(R/r_0)^2 - 1}, \quad \text{if } \alpha_0 = 0. \tag{3.135}$$

Now let $\sin\alpha_1 > 0$, (Figure 3.20k),

$$r_0d = \frac{R/r_0 \sin\alpha_0}{(R/r_0)^2 - 1} \tag{3.136}$$

For the limiting value of $\sin\alpha_0 = 1$ the limits of profile curve existence in the plane $\sin\alpha_1 = 0$ can be calculated (curve *uv*, Figure 3.26):

$$r_0d = \frac{R_0/r_0}{(R/r_0)^2 - 1}. \tag{3.137}$$

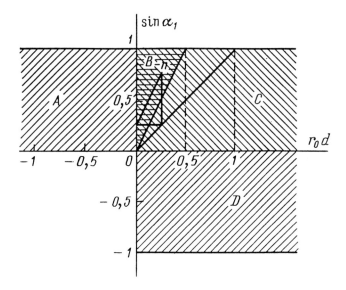

Fig. 3.27.   Section of the diagram of Figure 3.26 by the plane $R/r_0 = 1$, A – concave profile curves, B – convexo-concave profile curves, C-D – convex profile curves.

In the plane $\sin \alpha_1 = 1$

$$r_0 d = \frac{1 - (R/r_0) \sin \alpha_0}{1 - (R/r_0)^2} \tag{3.138}$$

For the limiting values of $\sin \alpha_0 = 1$ and $\sin \alpha_0 = 0$ the following equations are obtained respectively for curve EQ (Figure 3.26)

$$r_0 d = \frac{1}{1 + R/r_0} \tag{3.139}$$

and for curve HT (Figure 3.26)

$$r_0 d = \frac{1}{1 - R/r_0}. \tag{3.140}$$

The latter case can only exist for negative pressures. In the plane $R/r_0 = 1$ the range of profile curve existence is defined by the relation $\sin \alpha_1 = \sin \alpha_0$, which is also obvious for the reasons of symmetry.

Heights and shapes of convex curves. Application of the procedure described in the previous section allows expression of the shape and the height of the limiting profile curves ($\alpha_0 = \pi/2$) in terms of the Legendre functions:

$$z(r) = \frac{R_0 d - 1}{d}[F(k_4, \varphi_{10}) - F(k_4, \varphi_{11})]$$

$$- R_0[E(k_4, \varphi_{10}) - E(k_4, \varphi_{11})] \tag{3.141}$$

where

$$k_4^2 = (2R_0d - 1)/(1 + R_0d)^2, \quad \varphi_{11} = \arcsin \sqrt{d^2(R_0^2 - r^2)/(2R_0d - 1)}.$$

The expression for φ_{10} will be obtained from φ_{11} by replacing r with r_0; $z(R_0)$ can be obtained from (3.141) by replacing r with R_0.

For nonlimiting profile curves the procedure described in the previous case can be applied. The nonlimiting profile curve will be considered as a portion of the limiting profile curve. Then, from the values of the pressure d and the angle made by the liquid phase in contact with the growing crystal $\alpha_0 \mid_{r=R}$ specified for the nonlimiting profile curve, the value of R_0 for the corresponding limiting profile curve can be calculated from the expression for the first integral:

$$R_0 = \frac{1}{2d}(1 + \sqrt{1 - 4Rd \sin \alpha_0 + 4R^2 d^2}) \tag{3.142}$$

A nonlimiting profile curve will be a portion of this limiting profile curve limited by the abscissa $r = R$.

Heights and shapes of concave and convex-concave profile curves. The arbitrary constant of the first integral for limiting concave or convex-concave profile curves ($\alpha_0 = 0$) takes the form of $C_1 = -R_0^2 d$. The condition of problem solvability, $\Lambda^2 \le r^2 \le M^2$, leads to the necessity of satisfying the following two inequalities:

$$\Lambda^2 \le r_0^2, R_0^2 \le M^2. \tag{3.143}$$

The first condition coincides with the condition (3.140) that limits the peak value of the negative pressure. The second condition always holds:

$$1 + \sqrt{1 + 4R_0^2 d} > 0.$$

Substitution of (3.122) gives the shape of the profile curve in the form of the Legendre elliptical function of the first and second kinds.

$$z(r) = \frac{R_0^2}{M}[F(k_M, \varphi_x) - F(k_M, \varphi_\Lambda)]$$
$$- M[E(k_M, \varphi_x) - E(k_M, \varphi_\Lambda)], \tag{3.144}$$

Here the modulus k_M and the amplitudes φ_M and φ_Λ have the following form:

$$k_M^2 = \frac{M^2 - \Lambda^2}{M^2} = \frac{2\sqrt{1 + 4R_0^2 d^2}}{1 + 2R_0^2 d^2 + \sqrt{4R_0^2 d^2}}$$

$$\varphi_x = \arcsin[x(r)] = \arcsin \sqrt{\frac{M^2 - r^2}{M^2 - \Lambda^2}}$$

$$\varphi_\Lambda = \arcsin[x(R_0)] = \arcsin \sqrt{\frac{M^2 - R_0^2}{M^2 - \Lambda^2}}$$

The height of a profile curve is obtained from (3.144) by replacing r with r_0 in φ_x.

As was mentioned above, the nonlimiting profile curve can be reasonably represented as a portion of the limiting profile curve for which R_0 can be calculated from the expression for the first integral:

$$R \sin \alpha_0 - R^2 d = -R_0^2 d.$$

3.6.4. *Heights and Shapes of Profile Curves Possessing Ambiguous Projections on the Abscissa-Axis*

These curves are of interest because an increase in the melt column height corresponds to them. It is especially important for melt columns of nearly cylindrical shapes. Such profile curves can be analyzed in the following way.

Its lower portion (Figure 3.28) is a convex limiting profile curve with $R_0/r_0 > 1$, $\alpha_0 = \pi/2$. Its upper portion is a convex or a convex-concave profile curve for which the equatorial section of a drop of the radius equal to R_0 serves as a shaper. In these two cases on the basis of the expression for the first integral the relations between the dimensions of the shaper, r_0, of the meniscus, R_0, and the abscissa of the limiting profile curve peak, R_{01} (convex-concave), R_{02} (convex) is obtained:

$$r_0 \sin \alpha_1 - r_0^2 d = R_0 - R_0^2 d, \tag{3.145}$$

$$R_0^2 d - R_0 = -R_{01} + R_{01}^2 d = R_{02}^2 d \tag{3.146}$$

Note that since the gravitational force is disregarded the meniscus should be symmetrical relative to the point with the abscissa R_0 (Figure 3.28). Hence, if the lower portion of the meniscus from r_0 to R_0 is plotted, the upper one from R_0 to r_0 can be plotted symmetrically and only the meniscus portion $R < r_0$ requires further analysis. A few specific cases will be considered.

Drop squeezing out of the shaper. If it is assumed that $R_{02} = 0$, (3.45) and (3.46) represent the parameter relations for a symmetric drop squeezed out of the shaper with the diameter of $2r_0$:

$$\sin \alpha_1 = r_0 d, \ R_0 = 1/d. \tag{3.147}$$

The upper part of the drop is convex-concave. This case is possible only if $R_{01} < r_0$, $2r_0 d > \sin \alpha_1 > r_0 d$ as only then the following inequality holds:

$$r_0^2 d - r_0 \sin \alpha_1 = R_0^2 d - R_0 = R_{01}^2 d - R_{01}. \tag{3.148}$$

Then

$$R_{01} = \frac{1}{2d}(1 - \sqrt{1 + 4r_0^2 d^2 - 4r_0 d \sin \alpha_1})$$

$$R_0 = \frac{1}{2d}(1 + \sqrt{1 + 4r_0^2 d^2 - 4r_0 d \sin \alpha_1})$$

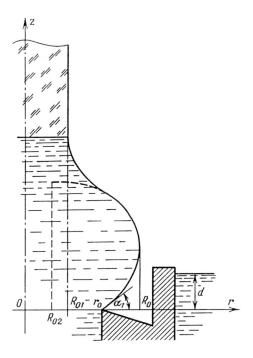

Fig. 3.28. Crystal growth for the profile curve with an ambiguous projection on the abscissa-axis.

$$\sin \alpha_1 = \frac{R_{02}}{r_0} + r_0 d[1 - (R_{01}/r_0)^2].$$

The height of the *limiting profile curve* can be written down as the sum of the heights of the upper and lower portions of the profile curves.

On the basis of our previous calculations, the height of a two-valued convexo-concave profile curve can be obtained in the form of the Legendre elliptical functions of the first and second kinds

$$z = \frac{R_0 d - 1}{d} F(k_4, \varphi_{10}) - R_0 E(k_4, \varphi_{10})$$

$$R_{01} \left[F(k_5, \frac{\pi}{2} - F(k_5, \varphi_{12})) \right]$$

$$+ \frac{1 + R_{01}d}{d} \left[E(k_5, \frac{\pi}{2}) - E(k_5, \varphi_{12}) \right] \qquad (3.149)$$

where

$$k_5^2 = \frac{1 - 2R_{01}d}{(1 - R_{01}d)^2}, \quad \varphi_{12} = \arcsin \sqrt{1 - \frac{d^2(R_0^2 - R_{01}^2)}{1 - 2R_{01}d}}.$$

The height of nonlimiting profile curves can be calculated from formula (3.149) but

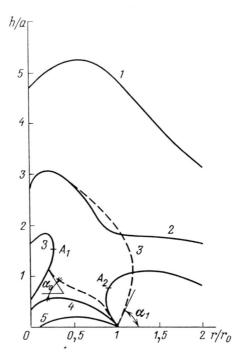

Fig. 3.29.   Boundary curves of two-valued (1, 2, 3) and one-valued (3, 4, 5) meniscus for $\sin \alpha_0 = 0.9$ and various $r_0 d$-values; 1 and 4 – 0.4; 2 – 0.6; 3 – 0.85; 5 – (-0.4); the dashed curves are profile ones.

$$\varphi_{13} = \arcsin \sqrt{1 - \frac{d^2(R^2 - R_{01}^2)}{1 - 2R_{01}d}}.$$

should be used as the argument of the elliptical function instead of $\pi/2$.

Figure 3.29 shows calculated for $\sin \alpha_0 = 0.9$ limiting curves characterizing height variations for one- and two-valued menisci depending on R/r_0 for various values of $r_0 d$. Curves 1, 2 and the portions of Curve 3 above the points $A_1(R/r_0 = 0.3)$ and $A_2(R/r_0 = 0.82)$ correspond to two-valued menisci. Curves $4,5$ and the lower portions of Curve 3 correspond to one-valued menisci. There is a break in the curves between the points A_1 and A_2, i.e., menisci do not exist for some particular crystal dimensions and certain pressures. It is also demonstrated by the capillarity diagram (Figure 3.21).

The capillarity diagram section by the plane $\sin \alpha_1 = 1$ can be plotted (Figure 3.30). From Equation (3.104) the relation of the parameters $r_0 d$ and R/r_0, e.g., for $\sin \alpha_0 = 0.9$, $\sin \alpha_1 = 1$ can be found:

$$r_0 d = \frac{1 - 0.9R/r_0}{1 - (R/r_0)^2}. \tag{3.150}$$

The expression obtained is the equation of the curve of intersection of the surface

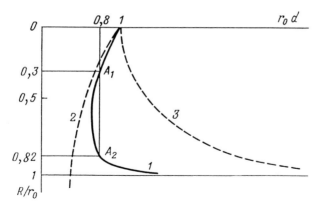

Fig. 3.30. Section of the space capillarity diagram by the plane $\sin \alpha_1 = 1$ line 1 is the trace of surface $\sin \alpha_0 = 0.9$; 2 – $\sin \alpha_0 = 1$; 3 – $\sin \alpha_0 = 0$.

P and the plane $\sin \alpha_1 = 1$ (Figure 3.22, curve kD). However, for the values of R/r_0 : $0.30 < R/r_0 < 0.82$ Equation (3.150) does not hold, i.e., existence of menisci with α_0 specified is impossible within this range.

3.7. Capillary Problem With the Gravity Force Allowed For

The surface of a liquid meniscus possessing axial symmetry with an allowance for the gravity force is described by Equation (1.12). Unlike Equations (1.13), (1.14), Equation (1.12) does not allow complete analysis and solution. For this reason numerical integration of this equation was carried out for a number of process parameter values that are of practical interest [87, 168].

A family of profile curves for $r_0 = 5.0$ and $d = 0.5$ is depicted in Figure 3.14 as an example. The curves have slopes towards the horizon equal to $15°, 20°, 30°, 45°, 60°, 75°, 85°$, respectively, at their contact points with the shaper.

Figure 3.31 shows the existence ranges of liquid columns (boundary curves) and partially the profile curves themselves for some practically interesting values of the problem parameters. In all the figures, the shaper free edges are brought into coincidence at the point of the coordinate origin to make comparison easier. The difference in slope towards the axis at their contact points with the shaper between adjacent curves is equal to $10°$.

Let us analyze the results obtained in detail. As was noted in the present chapter, liquid columns of three types can exist under capillary shaping conditions: convex, concave and convexo-concave. Numerical integration of Equation (1.12) allowing for the liquid column weight showed that all three types of liquid columns can really exist. The existence range of concave columns corresponds to the condition (3.11), while the existence ranges of convex and convexo-concave columns were calculated

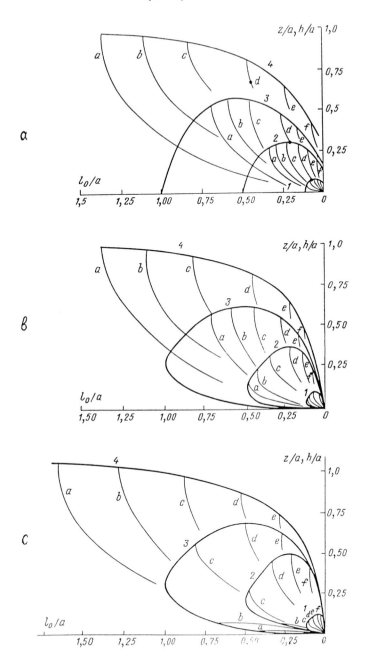

Fig. 3.31. Boundary curves for $\alpha_0 = 90°$ (1, 2, 3, 4) and some limiting profile curves (a, b, c, d, e, f) for various shaper radii: $1 - r_0 = 0.1$; $2 - r_0 = 0.5$; $3 - r_0 = 1.0$; $4 - r_0 = 5.0$; $r_0 d = 0$ (a); $r_0 d = 0.25$ (b); $r_0 d = 0.5$ (c). In Figure 3.31a the points show the positions of the boundary curves maxima for small Bond numbers.

by numerical integration of Equation (1.12). Figure 3.5 gives the summarized results. As can be seen from the figure, with r_0 increasing, the existence range of convex columns narrows diverging towards smaller α_0-values, the boundary between the ranges A and B coinciding with the corresponding boundary for the profile curves described by the capillary equation (1.14) (Figure 3.21) and the boundary between the ranges B and C coinciding with the corresponding boundary for $r_0 \rightarrow 0$. Figure 3.5 allows preliminary selection of the process parameters providing the desired liquid column shape without accurate calculations.

3.8. Calculation of the Transition of the Wetting Boundary Condition to the Catching Condition

Paragraph 3.1 qualitatively describes the transition from the catching boundary condition at the lower edge of the shaper to the wetting condition and then to the catching boundary condition at the upper free edge of the shaper with the pressure of the liquid fed into the shaper increasing. The results of the present paragraph allow the definition of quantitative characteristics of the above-described transition for a melt column possessing axial symmetry.

Let us formulate the problem. The shaper is made from a material whose wetting angle is Θ. The shaper has a hole with vertical walls. The hole diameter is $2r_0$. The pressure ranges satisfying the catching boundary condition at the lower free edge of the shaper, the wetting boundary condition on the hole walls and the catching boundary condition at the upper free edge of the shaper are to be defined.

In Figure 3.2a the pressure d is numerically equal to the distance between the liquid free surface level and the shaper lower free edge. The angle α_1 made by the line tangent to the liquid surface at the point of its contact with the shaper surface with the horizontal and the angle χ used to describe Figure 3.2a can be interrelated in the following way:

$$\alpha_1 = \chi - \pi/2.$$

In the diagram Figure 3.2a the catching boundary condition is satisfied at the shaper lower free edge where:

$$0 < \alpha_1 < \Theta - \pi/2, \tag{3.153}$$

the wetting boundary condition is satisfied if

$$\alpha_1 = \Theta - \pi/2, \tag{3.154}$$

the catching boundary condition is satisfied at the upper free edge of the shaper if

$$\Theta > \alpha_1 > \Theta - \pi/2 \tag{3.155}$$

Thus our problem is reduced to defining the relation between α_1-values and the pressures under which liquid is fed into the shaper hole.

Comparison of Figure 3.2 a and Figure 3.31 shows that the profile curves of Figure 3.2a are the limiting case of convex curve existence for which $\frac{dz}{dr}|_{R=0}=0$. Hence, the conditions of such meniscus existence coincide with the conditions of convex-to-convexo-concave profile curve transition. These conditions can be calculated from Figure 3.6 (the boundaries between ranges B and C) and the way of solving the problem formulated seems to be as follows.

Firstly, from Θ specified the α_1-value satisfying the wetting condition can be calculated:

$$\alpha_1 = \Theta - \pi/2$$

Then from α_1 calculated and r_0 specified, the desired pressure can be found using Figure 3.5. Among the notations of Figure 3.2a the pressure calculated corresponds to d_1. The catching boundary condition is satisfied at the shaper lower free edge where

$$0 \leq d \leq d_1 \tag{3.156}$$

The wetting boundary condition is satisfied on the hole walls where

$$d_1 \leq d \leq d_0 + d_1 \tag{3.157}$$

The catching boundary condition is satisfied at the shaper upper free edge where

$$d \geq d_0 + d_1 \tag{3.158}$$

Let the shaper have a hole in the form of a truncated cone and the free-edge angle be equal to B_0 (Figure 3.1c). The interrelation of the shaper geometric parameters is obvious from Figure 3.1c:

$$\operatorname{tg} B_0 = d_0/(r_H - r_0) \tag{3.159}$$

Firstly, the value of α_1 satisfying the wetting boundary condition on the hole walls can be defined from the values Θ and B_0 specified in the following way:

$$\alpha_1 = \Theta + B_0 - \pi. \tag{3.160}$$

Then from the values of α_1 calculated and r_H specified the pressure satisfying the wetting boundary condition in the vicinity of the lower free edge of the shaper can be defined using Figure 3.5. However, in contrast to the previous case when the value of d_1 remained constant with the shaper being immersed in the liquid, in this case the value of d_1 changes due to the varying diameter of the shaper. The behavior of d_1 can be seen in Figure 3.5 if a straight line parallel to the abscissa-axis is drawn on the level of $\sin \alpha_1$, the values of $r_H d_{10}$, $r_H d_{1H}$ are found and the corresponding pressures d_{1H}, d_{10} are calculated. The maximum d_1-value will be equal to d_{10} corresponding to the shaper free-edge diameter equal to $2r_0$.

So, the catching boundary condition is satisfied at the shaper lower free edge for

$$0 < d \leq d_{1H}. \tag{3.161}$$

The wetting boundary condition is satisfied on the hole walls for

$$d_{1H} < d < d_{10} + d_0 \tag{3.162}$$

The catching condition is satisfied at the upper free edge of the shaper if

$$d_{10} + d_0 \leq d. \tag{3.163}$$

Here, the following should be emphasized.

Firstly, all the quantities in the calculations discussed are measured in units of the capillary constant.

Secondly, in practical application of the results obtained the phenomenon of wetting angle hysteresis should be allowed for. If the shaper is immersed in the liquid the on-run wetting angle Θ_{on} should be taken into consideration; if the shaper is taken off the liquid the off-run wetting angle Θ_{off} should be taken into account.

Figure 3.2b shows that a similar transition of the boundary conditions will take place when the seed-to-shaper hole dimension ratio is changed. From the criteria (3.153) and (3.154) it follows that a change in the α_1-angle value should be achieved to provide the above-mentioned transition of the boundary conditions. As follows from Figure 3.31, such a change in the α_1-angle value can be obtained, firstly, by changing the pressure and, secondly, by changing the seed-to-shaper hole dimension ratio. As a result of numerical processing of the curves of the type shown in Figure 3.31 the angle α_1 was plotted as a function of the clearance between the seed or the cylindrical crystal being grown and the shaper hole $l_0 = r_0 - R_0$ for various pressures d under the condition of complete seed wetting by the melt or for *zero* growth angle ($\alpha_0 = \pi/2$) (Figure 3.32).

Then the problem is to be solved as follows. Firstly, from (3.153) the angle α_1 that provides the wetting condition should be calculated. Then from the values of α_1 calculated and r_0 specified and using the curves of Figure 3.32 (for intermediate r_0-values similar curves can be plotted by means of interpolation) the criteria of some certain boundary condition existence can be found. For this purpose, a straight line parallel to the axis of ordinates should be drawn from the abscissa point corresponding to the calculated value of α_1. The points where this straight line crosses curves 1–5 and the range to the left of this straight line correspond to the wetting condition, those to the right correspond to the catching condition at the free edge of the shaper, the pressure being counted off from the upper free edge of the shaper. Figure 3.33 shows the wetting-to-catching boundary condition transition for constant shaper dimensions with the pressure and the clearance between the shaper and the seed varying. The boundary curves were plotted using numerical calculation of Equation (1.12).

The above analysis is valid for profile curves described by Equation (1.14), the relations required can be derived analytically. Equation (3.104) can be used for this purpose. From (3.104) the angle α_1 is a function of α_0, d, R, r_0. Thus, by varying these parameters the required boundary condition on the shaper walls or at its free edges can be achieved using the procedure described above. In particular, the pressure satisfying the shaping conditions described can be defined:

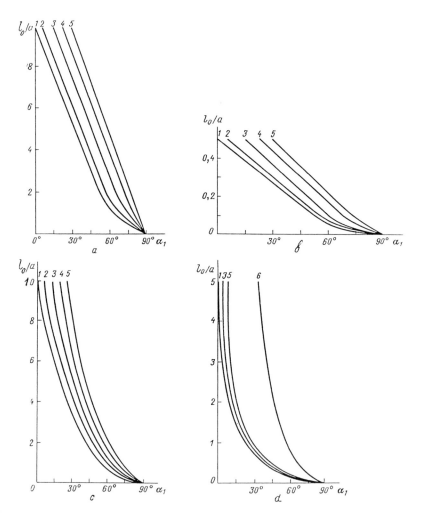

Fig. 3.32.   The values $l_0 = r_0 - R$ vs. the angle α_1, r_0 being equal to: 0.1 (a), 0.5 (b), 1.0 (c), 5.0 (d); for the curves (1–6) $r_0 d$ is as follows: 1 – 0; 2 – 0.1; 3 – 0.25; 4 – 0.4; 5 – 0.5; 6 – 2.5.

$$d = (r_0 \sin \alpha_1 - R \sin \alpha_0)/(r_0^2 - R^2)$$

With $\alpha_0 = 0$, the case shown in Figure 3.2a is obtained, i.e., drop squeezing out of the shaper. In this case $d = \sin \alpha_1 / r_0$ and the profile curve height can be calculated from the formula (3.129) rearranged using the previous relation

$$h = \frac{1}{2d} \arcsin(r_0 d)^2 = \frac{1}{2d} \arcsin(\sin \alpha_1)^2 \qquad (3.164)$$

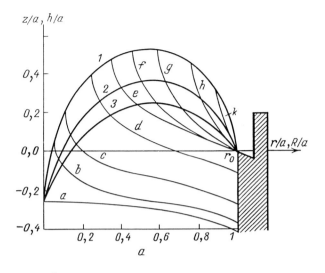

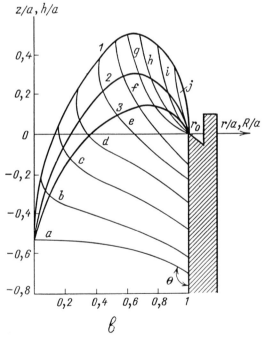

Fig. 3.33. Boundary curves (1–3) for the following α_0-values: $1 - \alpha_0 = 90°$; $2 - \alpha_0 = 60°$; $3 - \alpha_0 = 45°$ and profile curves $z(r)$ (a–j) for various seed-shaper dimension ratios corresponding to the transition of the wetting boundary condition (a–g) to the catching condition (g–j): $\alpha_1 = 15°$ (a), $\alpha_1 = 30°$ (b).

3.9. Influence of Pressure and Shaper-to-Crystal Clearance on the Shape of Profile Curves

The melt pressure and shaper-to-crystal clearance are the main parameters allowing control of the shape of profile curves. Here, numerically calculated data are given. Figure 3.34a shows the pressure influence on the existence ranges of limiting profile curves, Figure 3.34b depicts the same for nonlimiting profile curves. As the shaper-to-seed clearance increases, the heights of the profile curves first increase and then start decreasing with $\alpha_0 = const.$ Hence, for the given shaper diameter there exists some maximum profile curve height when the seed diameter is approximately *one-half* of the shaper diameter. If the shaper diameter $2r_0$ does not exceed *two* (the capillary constant a, is the unit of measurement of dimensionless quantities) and the value of r_0d does not exceed *one* (the case of particular practical interest), the order of magnitude of this maximum height lies within the limits of r_0 (Figure 3.35). For shaper holes whose diameters exceed 5 the maximum possible height of the profile curve lies within the limits of *one* (Figure 3.4, 3.31).

Figure 3.36 shows the pressure influence on the shape of profile curves, the growth angle and the crystal dimension being fixed. Figure 3.37 illustrates the pressure influence on the shape of profile curves for α_1-values exceeding $\pi/2$, i.e., two-valued menisci can be obtained.

3.10. Approximate Calculation of the Melt-Meniscus Height

In 3.2 the formula for the melt-meniscus height from [71] was given. As has already been mentioned, the formula cannot be used to calculate the melt-meniscus height for TPS without additional data.

The present chapter will obtain the data required [116]. The curves of Figure 3.32 avoid arbitrariness in boundary condition selection when using the formula. If pulling is carried out under the catching boundary condition, the angle α_1 can be determined from the pressure d known using the curves of Figure 3.32. If the wetting condition is satisfied the angle α_1 that can be calculated from (3.154) is known but the position of the melt column-shaper contact point is not known. Then using the curves of Figure 3.32 an inverse problem can be solved: the pressure d can be determined from the angle α_1 known.

Figure 3.5 can be used to define the melt meniscus type when the angle α_1 and the pressure d are known.

An approximate method of calculating the coordinates of the inflection point is proposed for convexo-concave curves. The ordinate of the inflection point can be represented in the form of its expansion into a Taylor series in the vicinity of the shaper free edge. Let r^* and z^* denote the coordinates of the inflection point and by restricting the series to one term of the expansion we obtain:

$$z^* = z_0'(r^* - r_0) \tag{3.165}$$

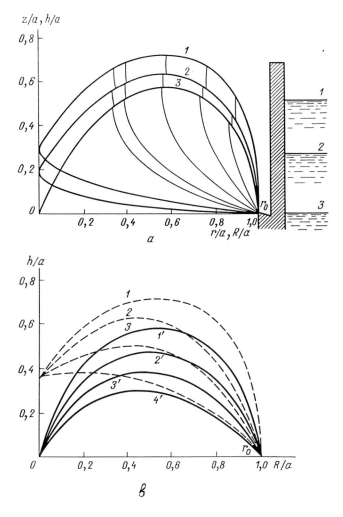

Fig. 3.34. Boundary $(1-3, 1'-4')$ and profile curves for a) various pressures $(r_0 = 1, \alpha_0 = \pi/2)$: $1 - d = 0.5$; $2 - d = 0.25$; $3 - d = 0$; and b) various angles α_0: 90^{circ} (curves $1, 1'$), $75°$ (curves $2, 2'$), $60°$ (curves $3, 3'$), $45°$ (curves $4, 4'$); and pressures d: 0.5 (curves 1, 2, 3, 4), 0 (curves $1', 2', 3', 4'$).

Here the coordinate z^* is measured off from the shaper level $(z^* > 0)$. At the inflection point $z'' = 0$ and Equation (1.12) takes the following form:

$$z' + 2(d - z^*)r^*(1 + z'^2)^{1/2} = 0. \tag{3.166}$$

Two equations (3.165) and (3.166) can be used to calculate the coordinates of the inflection point. From here it follows:

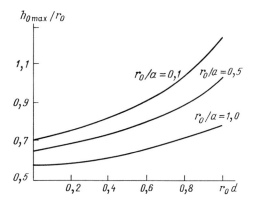

Fig. 3.35.   Relative maximum heights of the profile curves vs. $r_0 d$ for shapers of various dimensions r_0.

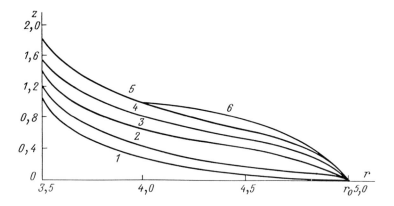

Fig. 3.36.   Profile curves (1–6) with the shaper dimension $r_0 = 5.0$, the crystal dimension $R = 3.5$ and the growth angle $\psi_0 = 10°$ under various pressures d: $1 - 0$; $2 - 0.2$; $3 - 0.4$; $4 - 0.6$; $5 - 0.8$; $6 - 1.0$.

$$r^* = \frac{d + r_0 z_0'}{2 z_0} \pm \sqrt{\left(\frac{d + r_0 z_0'}{2 z_0'}\right)^2 + \frac{1}{2(1 + z_0'^2)^{1/2}}}$$

$$z'' = z_0'\left[\frac{d + r_0 z_0'}{2 z_0} \pm \sqrt{\left(\frac{d + r_0 z_0'}{2 z_0'}\right)^2 + \frac{1}{2(1 + z_0'^2)^{1/2}}} - r_0\right] \qquad (3.167)$$

The closer the inflection point to the origin of coordinates, the better the agreement between the inflection point coordinates calculated from (3.167) and the true ones. Figure 3.38 shows the positions of the inflection points calculated numerically and the ones calculated from (3.167). For curve *3* the inflection point coincides with the origin of coordinates. For all the curves $r_0 = 1.0$, $d_0 = 0.5$. If the accuracy obtained

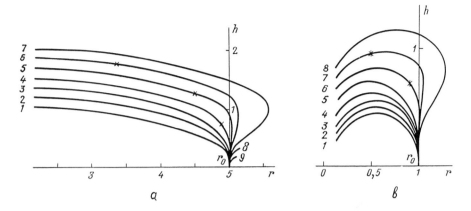

Fig. 3.37. Boundary curves (1–8) with $\psi_0 = 10°$, $r_0 = 5.0$ (a), $r_0 = 1.0$ (b) for various pressure-values d: 1 – 0; 2 – 0.2; 3 – 0.4; 4 – 0.6; 5 – 0.8; 6 – 1.0; 7 – 1.2; 8 – 1.4 (* the limits of two-valued meniscus existence range).

in this approximation is not sufficient two terms of z^* expansion into a Taylor series in terms of $(r^* - r_0)$ can be left. Then instead of (3.165) the following equation is obtained:

$$z^* = z_0''(r^* - r_0) + \frac{1}{2}z_0''(r^* - r_0)^2 \qquad (3.168)$$

Here

$$z_0'' = -\frac{z_0'}{r_0}(1 + z_0'^2) - 2d(1 + z_0'^2)^{3/2} \qquad (3.169)$$

Simultaneous solution of (3.166) and (3.168) gives a cubic equation for $r*$.

3.11. Limits of Capillary Equation Applicability Depending Upon the Bond Number Value

General considerations concerning the necessity of taking account of the influence of the capillary and gravity forces on melt meniscus shaping were stated in 1.2. Here quantitative relations will be derived, i.e., the limits of applicability of Equations (1.13) and (1.14) instead of Equation (1.12) will be defined.

Large Bond numbers. The range of Equation (1.13) applicability will be defined. Figure 3.4 gives a family of profile curves for $r_0 = 5$ and $2r_0d = 5$ plotted by solving Equation (1.12). In that figure, crosses show the heights of the profile curves calculated from Equation (1.13) for the same boundary conditions. In the range of α_1-values of 75° to 15° the difference in the maximum heights of the profile curves calculated from Equation (1.12) and Equation (1.13) is (1–8)%, respectively. Such

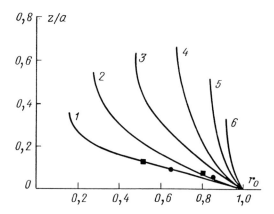

Fig. 3.38. Positions of □ – actual and ● – calculated from (3.167) points of inflection on the profile curves.

accuracy is quite sufficient for practical application especially because the pulling conditions corresponding to $\alpha_1 = 15^0$ are so seldom selected. Such conditions are close to pulling crystals by the Czochralski technique and, as will be shown below, the profile curve corresponding to such a case does not provide proper control of the geometrical dimensions of products. Usually, $\alpha_1 > 15°$ and application of Equation (1.13) for $r_0 > 5$ provides the accuracy of profile curve height calculation exceeding 8%.

Small Bond numbers. The position of the profile-curve maximum for Equation (1.14) with no pressure applied can be defined from (3.92). In Figure 3.31a the results obtained using this formula and the data calculated for $r_0 = 0.1$, $r_0 = 0.5$, $r_0 = 1.0$ are compared. For $r_0 = 0.5$ the difference in results does not exceed 8%. Therefore, in practice, Equation (1.14) with $2r_0 < 1$ can be used to calculate the heights of profile curves.

3.12. The Capillary Coefficients A_{RR} and A_{Rh}

To define the character of TPS growth depending on the values of the system pa-rameters, as follows from Equations (2.3) and (2.4), the behavior of $\partial\alpha_0/\partial R$ and $\partial\alpha_0/\partial h$ should be analyzed [77, 78]. With this, the following parameters can be used as variables: firstly, the boundary condition specified by the shaper (catching or wetting); secondly, the pressure for the catching boundary condition (it is controlled by changing the shaper position relative to the level of the melt free surface) or the relation between the growth angle and the shaper wall angle for the wetting boundary condition (it is controlled either by the wall slope or by selecting a proper material for the shaper); thirdly, the relation between the cross-section dimensions of the crystal being grown and the shaping contour. The meniscus is assumed to make an angle equal to the growth angle at the crystal free edge.

3.12.1. *Small Bond Number, the Catching Boundary Condition, Zero Pressure*

In this case, not only can the sign of the derivatives $\partial\alpha_0/\partial R$ and $\partial\alpha_0/\partial h$ be defined but also an analytical expression for them can be derived. The boundary curve equation (3.90) can be represented in the following form:

$$F(h, R, \alpha_0) = R\sin\alpha\left(\operatorname{arch}\frac{r_0}{R\sin\alpha_0} - \operatorname{arch}\frac{1}{\sin\alpha_0}\right) - h = 0. \qquad (3.170)$$

The expressions for $\partial\alpha_0/\partial R$ and $\partial\alpha_0/\partial h$ can be written down:

$$\frac{\partial\alpha_0}{\partial R} = -\frac{F_R'}{F_{\alpha_0}'}, \quad \frac{\partial\alpha_0}{\partial h} = -\frac{F_h'}{F_{\alpha_0}'}, \quad F_h' = -1,$$

$$F_R' = \sin\alpha_0\left(\operatorname{arch}\frac{r_0}{R\sin\alpha} - \operatorname{arch}\frac{1}{\sin\alpha}\right) - \frac{r_0\sin\alpha_0}{\sqrt{r_0^2 - R^2\sin^2\alpha_0}} \qquad (3.171)$$

$$F_{\alpha_0}' = R\cos\alpha_0\left(\operatorname{arch}\frac{r_0}{R\sin\alpha} - \operatorname{arch}\frac{1}{\sin\alpha}\right)$$
$$-\frac{Rr_0\cos\alpha_0}{\sqrt{r_0^2 - R^2\sin^2\alpha_0}} + R \qquad (3.172)$$

Figure 3.39 shows the behavior of the functions $\partial\alpha_0/\partial R$ and $\partial\alpha_0/\partial h$ in the range of $0 \le R \le r_0$ for two values of $\alpha_0(\alpha_0^{(1)} > \alpha_0^{(2)})$.

3.12.2. *Small Bond Numbers, the Catching Boundary Condition, Non-Zero Pressure*

For this case, the functional relations of the parameters h, R and α_0 are specified by Equation (3.120) where the expression for the arbitrary constant C_1 is derived from (3.103):

$$C_1 = R^2 d - R\sin\alpha_0, \qquad (3.173)$$

$$h(R) = \int_R^{r_0} \frac{r^2 d - R^2 d + R\sin\alpha_0}{\sqrt{r^2 - (r^2 d - R^2 d + R\sin\alpha_0)^2}} dr. \qquad (3.174)$$

The derivatives $\partial\alpha_0/\partial R$ and $\partial\alpha_0/\partial h$ can be calculated by direct differentiation of Equation (3.174):

$$\frac{\partial\alpha_0}{\partial h} = \{R\cos\alpha_0\int_R^{r_0} r^2[r^2 - (r^2 d - R^2 d + R\sin\alpha_0)^2]^{-3/2} dr\}^{-1} \qquad (3.175)$$

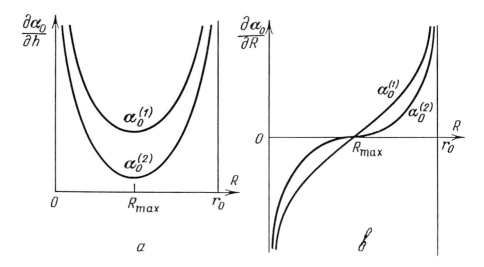

Fig. 3.39.   The functions $\partial\alpha_0/\partial h$ (a) and $\partial\alpha_0/\partial R$ (b) within the range of R-values from 0 to r_0.

$\frac{\partial\alpha_0}{\partial R} = -\frac{h'_R}{h'_{\alpha_0}}$, then $h'_{\alpha_0} = (\frac{\partial\alpha_0}{\partial h})^{-1}$ is prescribed by (3.175)

$$h'_R = -\mathrm{tg}\alpha_0 + (\sin\alpha_0 - 2r_0 d)$$

$$\times \int_R^{R_0} r^2[r^2 - (r^2 d - R^2 d + R\sin\alpha_0)^2]^{-3/2}dr \qquad (3.176)$$

From (3.175) it follows that $\partial\alpha_0/\partial h > 0$ within the whole range of $0 \leq R \leq r_0$, i.e., the behavior of this function is similar to its behavior in the case of $d = 0$. The behavior of the function $\partial\alpha_0/\partial R(R)$ depends on the relation between $\sin\alpha_0$ and $2r_0 d$.

For convex profile curves $\sin\alpha_0 < \sin\alpha_1$ if $R/r_0 < 1$. Since the condition of such curve existence (Figure 3.21) is $\sin\alpha_1 < 2r_0 d$, the condition $\sin\alpha_0 < 2r_0 d$ is satisfied within the whole range of R-values, the expression (3.176) is negative and hence $\partial\alpha_0/\partial R > 0$.

For concave profile curves $\sin\alpha_0 > \sin\alpha_1$ if $R/r_0 < 1$. Since the condition of such curve existence (Figure 3.21) is $\sin\alpha_1 > 2r_0 d$, the condition $\sin\alpha_0 > 2r_0 d$ is satisfied within the whole range of R-values, the sign of (3.176) depends on the values of the first and the second terms and the behavior of the function $\partial\alpha_0/\partial R(R)$ is similar to its behavior for $d = 0$ (Figure 3.39): $\partial\alpha_0/\partial R < 0$ for $R < R_m$, $\partial\alpha_0/\partial R(R) > 0$ for $R > R_m$, where R_m is the maximum point on the boundary curve $h(R)\,|_{\alpha_0=const}$ (Figure 3.25).

The behavior of the function $\partial\alpha_0/\partial R(R)$ for convexo-concave profile curves is the same.

3.12.3. *Small Bond Numbers, the Angle-Fixation Boundary Condition*

Two profile curves will be considered (Figure 3.40). From the first integral (3.173) for these profile curves it follows that

$$C_1 = r^2 d^{(1)} - r \sin \alpha_0^{(1)} = r_0^2 d^{(1)} - r_0 \sin \alpha_1 \tag{3.177}$$

$$C_2 = r^2 d^{(2)} - r \sin \alpha_0^{(2)} = r_0^2 d^{(2)} - r_0 \sin \alpha_1 \tag{3.178}$$

Let $d^{(1)} < d^{(2)}$, i.e., the first profile curve lies above the shaper wall and hence in the whole range of r-values the profile curves do not intersect (Figure 3.40a). At the crystal free edge the following conditions are satisfied:

$$R^2 d^{(1)} - R \sin \alpha_0^{(1)} = r_0^2 d^{(1)} - r_0 \sin \alpha_1$$

$$R^2 d^{(2)} - R \sin \alpha_0^{(2)} = r_0^2 d^{(2)} - r_0 \sin \alpha_1.$$

Hence

$$\sin \alpha_0^{(1)} - \sin \alpha_0^{(2)} = \frac{(r_0^2 - R^2)(d^{(2)} - d^{(1)})}{R} > 0$$

Since the profile curve characterized by $\alpha_0^{(1)}$ lies higher, i.e., a larger value of h corresponds to it, it follows that a positive increment in α_0 corresponds to a positive increment in h. Hence,

$$\partial \alpha_0 / \partial h > 0$$

The sign of the function $\partial \alpha_0 / \partial R$ is determined by the sign of h'_R as in (3.176). The following cases should be distinguished.

$\alpha_0 < \alpha_1$ (it should be remembered that $\alpha_0 = \pi/2 - \psi_0, \alpha_1 = \Theta - \pi/2$). The profile curves are either convex or convexo-concave in this case: $2r_0 d > \sin \alpha_1$ (Figure 3.21), hence $2r_0 d > \sin \alpha_0$, according to (3.176) $h'_R < 0$, whence it follows that $\partial \alpha_0 / \partial R > 0$ in the whole range of R-values.

$\alpha_0 > \alpha_1$. The profile curves are either concave or convexo-concave. The sign of h'_R depends on the relation of α_0, α_1, $r_0 d$, R/r_0. The most interesting case is $R/r_0 \rightarrow 1$. The range of convexo-concave column existence is very small (Figure 3.21) in this case, therefore, concave columns can exist in the main. In this case $2r_0 d < \sin \alpha_1$ and hence $2r_0 d < \sin \alpha_0$. For small clearances $r_0 - R$ the pressure is negative and therefore $h'_R > 0$ (Figure 3.40b), whence it follows that $\partial \alpha_0 / \partial R < 0$ for this case.

3.12.4. *Large Bond Number, the Catching Boundary Condition*

The functional relations between the parameters h, R and α_0 are specified by (3.43). In this, if the origin of coordinates is located at the shaper free edge the following expressions can be obtained for the integration constant A and the crystal-to-shaper clearance l_0:

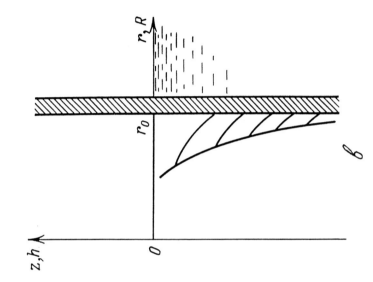

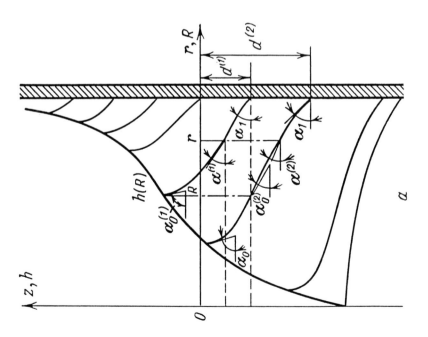

Fig. 40.   Profile $z(r)$ and boundary $h(r)$ curves (small Bond number) for wetting boundary conditions: $\alpha_0 > \alpha_1$ (a); $\alpha_0 < \alpha_1$ (b).

$$A = \cos\alpha_0 + (h - d)^2 = \cos\alpha_1 + d^2 \tag{3.179}$$

$$l_0(h) = \int_0^h \frac{\cos\alpha_0 + (h-d)^2 + (z-d)^2}{\sqrt{1 - [\cos\alpha_0 + (h-d)^2 - (z-d)^2]^2}} dz \tag{3.180}$$

hence

$$\frac{\partial\alpha_0}{\partial R} = \frac{1}{R'_{\alpha_0}} =$$

$$\left\{ \int_0^h \sin\alpha_0 \left\{1 - [\cos\alpha_0 + (h-d)^2 - (z-d)^2]^2\right\}^{-3/2} dz \right\}^{-1} \tag{3.181}$$

$$\frac{\partial\alpha_0}{\partial h} = -\frac{R'_h}{R'_{\alpha_0}}$$

$$R'_h = -\text{ctg}\alpha_0$$

$$+2\int_0^h (d-h)\{1 - [\cos\alpha_0 + (h-d)^2 - (z-d)^2]^2\}^{-3/2}\}dz \tag{3.182}$$

In this case $\partial\alpha_0/\partial R > 0$. The sign of the derivative $\partial\alpha_0/\partial h$ depends on the sign of $\partial h/\partial R$. For almost all the shaping conditions $\partial h/\partial R < 0$. The portion $0b$ in Figure 3.15 where $\partial h/\partial R > 0$ is an exception. This portion is characterized by the availability of menisci with ambiguous projections on the r-axis that can only exist under positive pressures. Thus, $\partial\alpha_0/\partial h > 0$ for almost all the shaping conditions with the exception of the portion $0b$ of Figure 3.15 where $\partial\alpha_0/\partial h < 0$.

3.12.5. *Large Bond Number, the Angle Fixation Boundary Condition*

Figure 3.13 shows the crystallization-front height $h|_{\alpha_0=const}$ versus the crystal-to-shaper clearance $l = r_0 - R$ for various relations between the growth angle and the wetting angle. The figures illustrate the following.

If $\alpha_1 > \alpha_0$ (it should be remembered that $\alpha_1 = \Theta - \pi/2, \alpha_0 = \pi/2 - \psi_0$), the following relations are obtained: $\partial\alpha_0/\partial R > 0, \partial\alpha_0/\partial h > 0$. Hence,

$$A_{RR} = -V\frac{\partial\alpha_0}{\partial R} < 0, \quad A_{Rh} = -V\frac{\partial\alpha_0}{\partial h} < 0.$$

If $\alpha_1 < \alpha_0$, as in the previous case $A_{Rh} < 0$ and the coefficient A_{RR} can be of both the signs. For small clearances $(l_0 < m)A_{RR} > 0$, for large clearances $(l_0 > m)A_{RR} < 0$ (Figure 3.13).

The results of the analysis of the capillary coefficients A_{RR} and A_{Rh} for all the cases considered are given in Table 3.1.

Conditions of crystallization				A_{RR}	A_{Rh}
Czochralski *technique*					

Capillary shaping technique	Small Bond number	Catching boundary condition	$\sin \alpha_0 < 2 r_0 d$	< 0	< 0
			$\sin \alpha_0 > 2 r_0 d$		
		Angle fixation boundary condition	$\alpha_0 < \alpha_1$	< 0	
			$\alpha_0 < \alpha_1$	> 0	
	Large Bond number	Catching boundary condition	single and double valued meniscus	< 0	
			double-valued meniscus $R \cong r_0$		> 0
		Angle fixation boundary condition	$\alpha_0 < \alpha_1$	< 0	< 0
			$\alpha_0 > \alpha_1$ $r - R > m$	< 0	< 0
			$\alpha_0 > \alpha_1$ $r - R > m$	> 0	

TABLE 3.1.

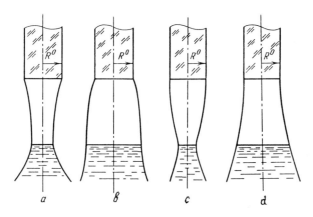

Fig. 3.41. Contracting (a, c) and widening (b, d) crystals for the cases of capillary stability (a, b) and instability (c, d).

3.13. Capillary Stability

It should be remembered (see 2.2.7) that the condition $A_{RR} < 0$ is a condition necessary to attain capillary stability. In this case $\delta R / R < 0$ and any crystal dimension perturbations will attenuate. It occurs in the following way.

Let us introduce the angle $\psi = \pi/2 - \alpha_0$ between the vertical and the line tangent to the melt at the crystal free edge (Figure 1.8). With $\psi = \psi_0$, a crystal of constant cross-section grows.

A change in crystal dimensions. With $A < 0$, a change in crystal dimensions leads to such change of the angle ψ that the crystal returns to the undisturbed state (Figure 3.19, point A). In the opposite case ($A_{RR} > 0$) a change in crystal dimensions results in further perturbation increase (Figure 3.19, point B).

Crystal-front displacement. In this case the angle ψ also changes departing from ψ_0. The crystal dimension will start changing and in case capillary stability is available a new stable state will be achieved: a crystal of constant cross-section of slightly different dimension will be pulled (Figure 3.19, points A' and A''). In case capillary stability is not available the crystal dimension will either increase or decrease without returning to the stable state (Figure 3.19, points B' and B''). The crystal transition regions for capillary stable and unstable growth are given in Figure 3.41 as an illustration.

Analysis of the data of Table 3.1 shows that the stability range is determined by the character of the boundary conditions and the shaper-to-crystal clearance value. The catching boundary condition and a small shaper to crystal clearance are preferable. Now it is possible to answer the question where the Czochralski technique ends and TPS begins. The *Czochralski technique differs from TPS in the absence of ranges of capillary stability.* Hence in such cases when shaper presence leads to capillary stability we deal with TPS. Thus, when pulling a circular cylindrical rod with the catching boundary condition satisfied the range of capillary stability exists only in case the crystal-to-shaper dimension ratio $R/r_0 \geq 1/2$. This ratio is the condition of the Czochralski-to-TPS growth transition.

3.14. The Conditions for Crystallization Stability

Formulation of the heat problem for TPS does not differ from that of the Czochralski technique. The only difference is that the one-dimensional heat-conductivity equation can more reasonably be applied to TPS since the crystals grown have smaller cross-sections and the shape of the melt column is closer to the cylindrical one.

Now crystallization process stability of TPS can be investigated by analyzing the conditions that satisfy the inequalities (2.10), (2.11). The capillary coefficients A_{RR} and A_{Rh} are given in Table 3.1. Analysis of the results of the heat problem solution (2.42)–(2.63) shows that $A_{hR} > 0$ and the conditions that make $A_{hh} < 0$ can be provided.

When thin rod-shaped crystals are pulled (small Bond number) when the catching condition is satisfied at the shaper sharp edge, interstabilization of the parameters takes place in the system $(A_{Rh}A_{hR} < 0)$. In this case stable growth is observed in the range of $R_b > R_m$, where R_b is the stability limit calculated from the conditions (2.10), (2.11). In this case $R_b > R_m$ (Table 3.1) for supercooled melts and $R_b \leq R_m$ for overheated ones. In other words, when $A_{hh} < 0$ the range of overall stability is wider than that of capillary stability. With capillary stability available, supercooled-melt stable growth is possible $(A_{hh} > 0)$, which allows pulling rate increase due to crystallization heat removal from the front not only through the crystal but also through the supercooled melt.

When the angle-fixation boundary condition is satisfied on the shaper wall, the availability of capillary stability depends on the growth angle-wetting angle relation. If the shaper is absolutely nonwettable $(\Theta = 180x, \alpha_1 = \pi/2)$ capillary stability can exist for small growth angles $(\psi_0 > 0, \alpha_0 < \alpha_1 = \pi/2)$. Otherwise, stable growth is possible only for $A_{hh} < 0$ if inequalities (2.10) and (2.11) hold.

While growing crystals in the shape of a plate or a cylinder of a large diameter (large Bond number) using the catching boundary condition, capillary stability is always present $(A_{RR} < 0)$. Besides, for crystal, whose dimensions lie outside the interval $0b$ (Figure 3.15), interstabilization exists $(A_{Rh}A_{hR} < 0)$. In this case stable growth is possible both for $A_{hh} < 0$ and for $A_{hh} > 0$ if the conditions (2.10) and (2.11) are satisfied.

For crystals whose dimensions lie within the interval $0b$, interstabilization cannot exist and stable growth is possible only from overheated melt if the condition (2.11) is satisfied.

When the angle-fixation boundary condition is satisfied for large Bond number, capillary stability can exist under the condition $\alpha_0 < \alpha_1$ (the growth angle ψ_0 should be larger than the complement of the wetting angle Θ). With $\alpha_0 > \alpha_1$, capillary stability is possible only for large crystal-to-shaper wall clearances $(r_0 - R > m)$.

With the clearance and the crystal dimension decreasing ($r - R < m$, $R/a \cong 1$), the coefficient A_{RR} becomes positive and capillary stability is absent.

The possibility of using capillary stability is an important advantage of TPS over the Czochralski technique. In this case, both overheated- and supercooled-melt stable growths of crystals of constant cross-sections are possible without application of automatic-control systems.

To numerically estimate the degree of stability a quantity inverse to the characteristic time of perturbation relaxation should be introduced.

With stability present, roots S of the characteristic equation (1.4) have negative real parts. Then $\min\{|\text{Re}S_1|, |\text{Re}S_2|\}$, where $|\text{Re}S_i|$ is the real part of the root S_i of the characteristic equation ($i = 1, 2$), will be regarded as a degree of stability S.

As was mentioned above, the degree of crystallization process stability can be improved and the range of stable growth can be extended by applying forced blowing in the zone above the crystallization front and by increasing the degree of melt overheating as well as by decreasing the pressure d in the meniscus.

From (2.48) it is obvious that the degree of heat stability (the modulus of the coefficient A_{hh}) is higher for the substances possessing higher thermal conductivity coefficients and lower latent melting heat per unit substance volume.

3.15. Stationary Values of R^0 and h^0 as Simultaneous Solutions of the Capillary and the Heat Problems

It has already been mentioned that to find stationary values of the crystal dimension R^0 and the meniscus height h^0 the function $h(R)$ calculated while solving the capillary and the heat problems should be simultaneously analyzed. The case of overheated-melt growth of a thin rod-shaper crystal with the catching boundary condition satisfied under zero pressure will be used as an example of this procedure in TPS.

In this case the function $h(R)$ calculated while solving the capillary problem is described by expression (3.90), with $\alpha_0 = \alpha_e$.

As a result of the heat-problem solution, the function $h(R)$ can be represented in the following way. For small pulling rates heat transfer caused by crystal and melt movement can be neglected. For this purpose the inequality $V^2/4k_i^2 \ll 2\mu_i/\lambda_i R$ should be satisfied. For $\zeta_S L \gg 1$, the crystal can be considered as an infinitely long one. For $\zeta_L L \ll 1$ the heat removal from the liquid-column lateral surface can be neglected. Under such constraints imposed for the temperature gradients G_L and G_S the following estimates are obtained:

$$G_L \approx \frac{T_m - T_0}{h}, \qquad G_S = -(T_0 - T_e)\sqrt{\frac{2\mu_S}{\lambda_S R}}$$

Then the equation of crystallization-front heat balance can be written as follows:

$$\sqrt{\frac{\beta}{R}} - \frac{1}{h} = \eta V. \tag{3.183}$$

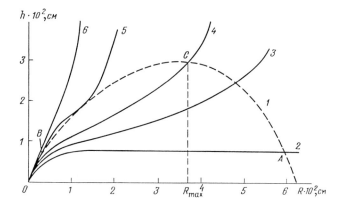

Fig. 3.42. Crystallization front height vs. crystal dimension for cylindrical silicon crystals from (1) – the capillary and (2–6) – the heat problem solution for various growth rates V_i where $V_{i+1} > V_i$ and $i = 2, 3, \ldots, 6$.

Here

$$\beta = \frac{2\mu_S \lambda_S^2 (T_0 - T_e)^2}{\lambda_L^2 (T_m - T_o)^2}, \qquad \eta = \frac{\mathcal{L}}{\lambda_L (T_m - T_0)}$$

Figure 3.42 shows the function $h(R)$ calculated from (3.90) (Curve 1) for $r_0 = 6 \cdot 10^{-2}$cm, $\psi_0 = 11°$ (silicon) and from (3.183) (curves 2-6) for five various pulling rates V. To plot the function $h(R)$ from (3.183) the values characteristic of silicon growth were taken, i.e., $\beta = 5 \cdot 10^4$ m^{-1}, $\eta = 10^7$s· m^{-2}.

The maximum possible crystal dimensions for fixed β- and η-values correspond to simultaneous solutions of the capillary and heat problems with $V = 0$ (point A, Figure 3.42). Depending on the parameter values cases of two, one or no simultaneous solutions exist. When two solutions exist the one corresponding to the smaller radius (point B) is always unstable as the inequality (2.11) is violated. Indeed, the inequality (2.11) can be represented in the following form:

$$A_{hh} A_{Rh} \left[\left(\frac{dh}{dR}\right)_T - \left(\frac{dh}{dR}\right)_C \right] > 0. \tag{3.184}$$

Here $\left(\frac{dh}{dR}\right)_T$ and $\left(\frac{dh}{dR}\right)_C$ are derivatives of the function $h(R)$ from (3.183) and (3.90), respectively. It is obvious from Figure 3.42 that for the solution corresponding to the smaller radius $\left(\frac{dh}{dR}\right)_T - \left(\frac{dh}{dR}\right)_C < 0$ and inequality (2.11) is violated (instability of the saddle-type). With the pulling rate increasing, the simultaneous solutions close in and for some critical pulling rate V_{ex} only one simultaneous solution exists (curve 5), Figure 3.42) and for $V > V_{ex}$ simultaneous solutions do not exist (curve 6).

For the simultaneous solution corresponding to the larger radius (point C) the inequality (2.11) always holds. If this R-value lies within the range of capillary stability ($R > R_m$) it is always stable. If $R < R_m$, for the solution to be stable inequality (2.10) should be satisfied.

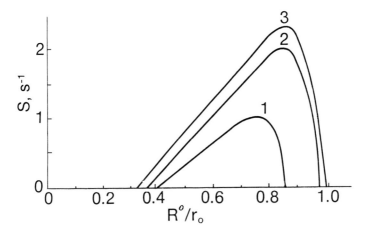

Fig. 3.43.   Stability vs. crystal-to-shaper dimension rate for: (1) – Al, (2) – Si, (3) – Al_2O_3.

Figure 3.43 shows the degree of stability S versus the value of the equilibrium crystal radius R^o. The right stability limit (the point corresponding to $S = 0$ in the range of large radii) satisfies the simultaneous solution with $V = 0$. The left stability limit satisfies the condition $A_{RR} + A_{hh} = 0$. From Figure 3.43 it follows that an optimal ratio of the dimension of the crystal being pulled R^0 to the shaper dimension r_0 (with the values of β and η fixed) when the maximum degree of stability can be achieved exists.

With the radius of the growing crystal increasing from *zero* to r_0, the type of stability changes in the following way: "unstable saddle" - "unstable node" - "unstable focus" - "stable focus" - "stable node" (Figure 2.1).

3.16. Growth of Tubular Crystals [123]

Cross-sections of tubular crystals are binary connected feasible areas and are characterized by two dimensions: the external radius of the tube R_1 and the inner one R_2. Thus, in this case a system with three degrees-of-freedom is obtained, i.e., in (1.1) $n = 3$, $X_1 = R_1$, $X_2 = R_2$, $X_3 = h$. Time variation of the quantities R_1 and h are described by (1.39) and (1.48) as before. The changing rate of both the inner tube radius R_2 and the external one R_1 can be defined from the condition of growth angle constancy.

$$\frac{dR}{dt} = V\,\mathrm{tg}(\alpha_{02} - \alpha_e) \tag{3.185}$$

The functional relation between the angle α_{02}, crystallization front position h and the inner tube radius R_2 can be found by solving the Laplace capillary equation for the internal meniscus part.

The linearized set of equations (1.3) characterizing the Lyapunov stability can in this case be represented in the following form:

$$\delta \dot{R}_1 = A_{R_1 R_1} \delta R_1 + A_{R_1 R_2} \delta R_2 + A_{R_1 h} \delta h$$

$$\delta \dot{R}_2 = A_{R_2 R_1} \delta R_1 + A_{R_2 R_2} \delta R_2 + A_{R_2 h} \delta h$$

$$\delta \dot{h} = A_{h R_1} \delta R_1 + A_{h R_2} \delta R_2 + A_{hh} \delta h \qquad (3.186)$$

From (1.46) and (3.185) for the capillary coefficients the following expressions can be obtained:

$$A_{R_1 R_1} = -V \frac{\partial \alpha_{01}}{\partial R_1}, \quad A_{R_1 R_2} = -V \frac{\partial \alpha_{01}}{\partial R_2}, \quad A_{R_1 h} = -V \frac{\partial \alpha_{01}}{\partial h},$$

$$A_{R_2 R_1} = V \frac{\partial \alpha_{02}}{\partial R_1}, \quad A_{R_2 R_2} = V \frac{\partial \alpha_{02}}{\partial R_2}, \quad A_{R_2 h} = V \frac{\partial \alpha_{02}}{\partial h}. \qquad (3.187)$$

Note that $A_{R_1 R_2} = A_{R_2 R_1} = 0$. This is to be connected, as the internal and the external parts of a meniscus are independent. The only thing connecting them is that in the Laplace capillary equation describing such meniscus the pressure should be the same if the internal and the external shaper free edges lie on the same level or the pressure difference equal to the difference in height between the free edges should be constant if they lie on various levels. The origin of coordinates should coincide with the free edges.

The heat coefficients can be calculated from (1.48):

$$A_{h R_1} = \frac{1}{\mathcal{L}} \frac{\partial G_{LS}}{\partial R_1}, \quad A_{h R_2} = \frac{1}{\mathcal{L}} \frac{\partial G_{LS}}{\partial R_2}, \quad A_{hh} = \frac{1}{\mathcal{L}} \frac{\partial G_{LS}}{\partial h} \qquad (3.187a)$$

Here the notation $G_{LS} = \lambda_L G_L - \lambda_S G_S$ is used. Characteristic cross-section dimensions R_1 and R_2 are included into the heat problem in the form of the perimeter to cross-section area ratio F. Thus $F = 2/R_1$ while crystallizing a rod of the radius R_1. While crystallizing tubes of not too large inner diameters heat is usually removed from the external surface of the tube only. In this case $F = 2R_1(R_1^2 - R_2^2)^{-1}$. If heat is also removed from the inner surface of the tube $F = 2(R_1 - R_2)^{-1}$. Then the coefficient (3.187a) can be represented in a more convenient form:

$$A_{h R_1} = \frac{1}{\mathcal{L}} \frac{\partial G_{LS}}{\partial F} \frac{\partial F}{\partial R_1} > 0,$$

$$A_{h R_2} = \frac{1}{\mathcal{L}} \frac{\partial G_{LS}}{\partial F} \frac{\partial F}{\partial R_2} < 0,$$

$$A_{hh} = \frac{1}{\mathcal{L}} \frac{\partial G_{LS}}{\partial h} < 0, \quad \text{if} \quad T_m > T_0. \qquad (3.188)$$

In case $F = 2(R_1 - R_2)^{-1}$, $A_{h R_1} = -A_{h R_2}$. For the set of equations (3.186) to be stable, according to the Routh-Gurvitz criterion it is necessary and sufficient that the following inequalities should hold:

$$A_{R_1R_1} + A_{R_2R_2} + A_{hh} < 0 \tag{3.189}$$

$$-A_{R_1R_1}A_{R_2R_2}A_{hh} + A_{R_1R_1}A_{R_2h}A_{hR_2} + A_{R_2R_2}A_{R_1h}A_{hR_1} > 0, \tag{3.190}$$

$$-(A_{R_1R_1} + A_{R_2R_2} + A_{hh})(A_{R_1R_1}A_{R_2R_2} + A_{R_1R_1}A_{hh} + A_{R_2R_2}A_{hh}$$
$$-A_{R_1h}A_{hR_1} - A_{R_2h}A_{hR_2}) - (A_{R_1R_1}A_{R_2R_2}A_{hh}$$
$$+A_{R_1R_1}A_{R_2h}A_{hR_2} + A_{R_2R_2}A_{R_1h}A_{hR_1}) > 0. \tag{3.191}$$

Let us analyze tube growth when the catching boundary condition is satisfied at the shaper sharp free edges with $d = 0$ (the melt free surface level in the crucible coincides with the level of the shaper sharp free edges). When analyzing stationary solutions of R_1, R_2 and h of the heat-balance equation for the crystallization front, (3.183) will be used. For $F = 2R_1(R_1^2 - R_2^2)^{-1}$ this equation will have the following form:

$$\sqrt{\frac{\beta R_1}{R_1^2 - R_2^2}} - \frac{1}{h} = \eta V \tag{3.192}$$

3.16.1. *Tubes of Large Diameters* $(r_1 \gg 1)$

(r_1 is the external and r_2 is the inner radii of the shaper). In this case from (3.43) with $d = 0$ it is obtained that

$$r_1 - R_1 = r_2 - R_2 = \int_0^h \frac{\cos\alpha_0 + h^2 - z^2}{\sqrt{1 - (\cos\alpha_0 + h^2 - z^2)^2}} dz \tag{3.193}$$

The heat-balance equation (3.192) can be represented in the following form:

$$h = \left[\sqrt{\frac{\beta R_1}{(r_1 + r_2)(2R_1 - r_1 - r_2)}} - \eta v\right]^{-1} \tag{3.194}$$

Here it is taken into account that $R_1 + R_2 = r_1 + r_2$ – it follows from (3.193). The qualitative behavior of the functions $h(r)$ from (3.193) and (3.194) is depicted in Figure 3.45. It is obvious that only one solution of the set of equations (3.193) and (3.194) exists and it is always stable. Indeed, from (3.193), (3.181) and (3.182) for the capillary coefficients (3.187) it follows that $A_{R_1R_1} < 0$, $A_{R_1h} < 0$, $A_{R_2R_2} < 0$, $A_{R_2h} > 0$. The signs of the heat coefficients are defined from inequalities (3.188). In this case inequalities (3.189)–(3.191) hold and the crystallization process is stable.

When allowing for the heat removal from the inner surface of the tube too, the case of plate growth discussed above is actually obtained.

3.16.2. *Tubes of Small Diameters* $(r_1 \ll 1)$

In this case the boundary problem to calculate the meniscus equilibrium shape includes Equation (3.82) and the following boundary conditions:

$$z\,|_{r=r_i}= 0, \tag{3.195}$$

$$z'\,|_{r=R_i}= \pm \mathrm{tg}\alpha_{0i} \tag{3.196}$$

where $i = 1, 2$. In (3.196) the plus sign corresponds to $i = 2$, and the minus sign corresponds to $i = 1$. By solving the boundary problem it is obtained from (3.86) that *for the external meniscus part*:

$$h(R_1) = R_1 \sin\alpha_{01}\left[\mathrm{arch}\,\frac{r_1}{R\sin\alpha_{01}} - \mathrm{arch}\,\frac{1}{\sin\alpha_{01}}\right]$$

$$0 \le R_1 \le r_1 \tag{3.197}$$

for the internal meniscus part: in the case of one-valued meniscus (the dashed line in Figure 3.44b)

$$h(R_2) = R_2 \sin\alpha_{02}\left[\mathrm{arch}\,\frac{1}{\sin\alpha_{02}} - \mathrm{arch}\,\frac{r_2}{R_2\sin\alpha_{02}}\right]$$

$$r_2 \le R_2' \le r_2/\sin\alpha_{02} \tag{3.198}$$

in the case of two-valued meniscus (the solid line in Figure 3.44b)

$$h(R_2) = R_2 \sin\alpha_{02}\left[\mathrm{arch}\,\frac{r_2}{R\sin\alpha_{02}} + \mathrm{arch}\,\frac{1}{\sin\alpha_{02}}\right]$$

$$0 \le R_2 \le r_2/\sin\alpha_{02} \tag{3.199}$$

The heat-balance equation can be represented in the form that follows from (3.183):

$$R_1(h) = \frac{\beta h^2}{2(\eta V h + 1)^2} + \sqrt{\frac{\beta^4 h^4}{4(\eta V h + 1)^4} + R_2^2(h)} \tag{3.200}$$

Substitution of $R_2(h)$ from (3.198) or from (3.199) into (3.200) and solution of the equation obtained simultaneously with (3.197) give the expressions for R_1, R_2 and h (Figure 3.46).

Cases when two, one or no solutions exist are possible depending on the values of the problem parameters.

To analyze the stability of the solutions obtained the signs of the coefficients A_{ij} included into the set of inequalities (3.189)–(3.191) should be defined. The coefficients $A_{R_1 R_1}$ and $A_{R_1 h}$ were analyzed before and their behavior within the interval $[0, r_1]$ is represented in Tables 3.1 and 3.2. The coefficients $A_{R_2 R_2}$ and $A_{R_1 h}$ can be easily found by differentiating (3.198) and (3.199). In this case if $h(R_2)$ is described by (3.198) $A_{R_2 R_2} < 0$ and $A_{R_2 h} > 0$. In case $h(R_2)$ is described by (3.199) the coefficients $A_{R_2 R_2}$ and $A_{R_2 h}$ are given in Table 3.2. The signs of the heat coefficients are defined from inequalities (3.188) and for a most simple heat problem formulation they are also given in Table 3.2.

Big Bond Number $(R_2 > a)\; d = 0$	Small Bond Number $(R_1 \ll a)\; d = 0$	
Negative	$O \quad A_{R_1 R_1}$ ⟶ R_1^* — R_1 — r_1 — R_1	$A_{R_1 R_1}$
Negative		$A_{R_1 h}$
Negative	$O \quad A_{R_2 R_2}$ ⟶ R_2^{**} — R_2^* — $\dfrac{r_2}{\sin\alpha_0}$ — R_2	$A_{R_2 R_2}$
Positive	$O \quad A_{R_2 h}$ ⟶ R_2^{**} — R_2	$A_{R_2 h}$
Positive		$A_{h R_1}$
Negative		$A_{h R_2}$
Negative, if $\;T_p \geq T_0$ Positive, if $\;T_p < T_0$		A_{hh}

TABLE 3.2.

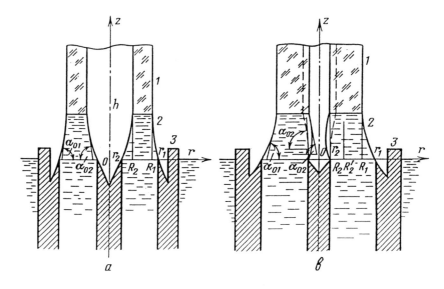

Fig. 3.44.   Growth of tubular profiles of large $r_2 > 1$ (a) and small $r_1 < 1$ (b) diameters.

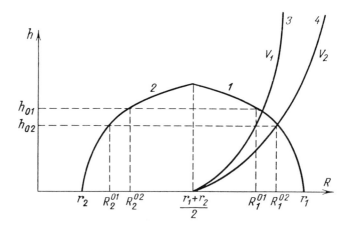

Fig. 3.45.   Simultaneous solution of the capillary (1, 2) and heat (3, 4) problems for tubes of large diameters: $1 - h(R_1)$ and $2 - h(R_2)$ from (3.193), 3, 4 $- h(R_1)$ from (3.194) for two various pulling rates V_1 and V_2 ($V_1 > V_2$).

The root of Type 1 (Figure 3.46) in the range where $h(R_2)$ is described by (3.198) is stable. Indeed, in this case the signs of the coefficients A are such that inequalities (3.189)–(3.191) are satisfied. The roots of Type 2 (Figure 3.46) for which $R_1 < R_2^*$ are unstable with $R_2^{**} < R_2$ (Table 3.2). Indeed, all the tree terms in (3.190) are negative and the inequality is violated. The roots of Type 2 (Figure 3.46) for which

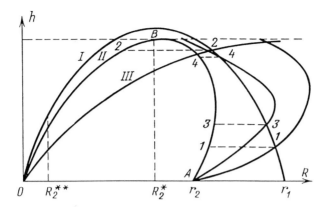

Fig. 3.46. Simultaneous solution of the capillary (1, 2) and heat (3, 4) problems for tubes of small diameter: curve $1 - _1(h)$ is plotted from (3.197), curve $2 - R_2(h)$: portion AB from (3.198), portion BO from (3.199), curves 3 and $4 - R_1(h)$ from (3.200) where the branch of curve 2 from 0 to R_2^* is used to plot curve 3, and the upper branch from R_2^* to r_2 is used to plot curve 4.

$R_2 < R_2^{**}$ are stable. While analyzing the stability of the roots of Types 3 and 4 (Figure 3.46) note that

$$A_{R_1 h} = -A_{R_1 R_1} \left(\frac{DR_1}{dh} \right)_C$$

where $\left(\frac{DR_1}{dh} \right)_C$ is a derivative of the function $R_1(h)$ from (3.197),

$$A_{R_2 h} = -A_{R_2 R_2} \left(\frac{DR_2}{dh} \right)_C,$$

where $\left(\frac{dR_2}{dh} \right)_C$ is a derivative of the function $R_2(h)$ from (3.199),

$$A_{hh} = -A_{h R_1} \left(\frac{DR_1}{dh} \right)_T,$$

where $\left(\frac{DR_1}{dh} \right)_T$ is a partial derivative of the function $R_1(h, R_2)$ from (3.200). Then (3.190) can be written in the following form:

$$A_{R_1 R_1} A_{R_2 R_2} A_{h R_1} \left[\left(\frac{dR_1}{dh} \right)_T - \left(\frac{dR_1}{dh} \right)_G \right] > 0 \qquad (3.201)$$

where

$$\left(\frac{dR_1}{dh} \right)_T = \left(\frac{\partial R_1}{\partial h} \right)_T + \left(\frac{\partial R_1}{\partial R_2} \right)_T \left(\frac{dR_2}{dh} \right)_C$$

is a total derivative of the function $R_2 = R_1[h, R_2(h)]$ from (3.200).

It is obvious that for a root of Type 4 (Figure 3.46) inequalities (3.201) and hence (3.190) are not satisfied, i.e., it is unstable. For a root of Type 3 (Figure 3.46) inequalities (3.201) and (3.190) are satisfied. For small growth angles ($\alpha_0 \approx \pi/2$) (the case most often used in practice) $|A_{R_1 R_1}| > |A_{R_2 R_2}|$ (to prove this statement it is sufficient to compare the derivatives $\frac{\partial \alpha_{01}}{\partial R_1}$ and $\frac{\partial \alpha_{02}}{\partial R_2}$). In this case (3.191) is also satisfied, i.e., a root of Type 3 is stable. In more general cases the problem of root 3 stability requires additional analysis.

Some limiting cases that are of practical interest should be analyzed.

3.16.3. Crystallization of Thick-Walled Tubes ($R_1 \gg R_2$)

In this case R_2^2 in Equation (3.192) can be disregarded and simultaneous solution of (3.192) and (3.197) allows the calculation of R_1 and h and then from (3.198) or (3.199) R_2 can be calculated. Since $A_{hR_2} = 0$ in this case, for the solution to be stable it is necessary and sufficient that the following inequalities should hold:

$$A_{R_2 R_2} < 0 \tag{3.202}$$

$$A_{R_1 R_1} A_{hh} < 0 \tag{3.203}$$

$$A_{R_1 R_1} A_{hh} - A_{R_1 h} A_{hR_1} > 0 \tag{3.204}$$

The stability of simultaneous solutions of R_1 and h from (3.192) and (3.197) that is determined by (3.203) and (3.204) has already been analyzed. Additionally, capillary stability of the inner radius (3.202) is required as its heat stabilization is absent ($A_{hR_1} = 0$). In this case, only one solution of type 2 and 3 exists (Figure 3.46), the roots of Type 2 being stable with $R_2 < R_2^{**}$ and unstable with $R_2^{**} < R_2 < R_2^*$ ((3.202) is violated), similar to the general case. The roots of Type 3 are stable as inequalities (3.202)–(3.204) are satisfied.

3.16.4. Crystallization of Thin-Walled Tubes ($R_1 - R_2 = \epsilon \ll R_2$)

Since the growth angle ψ_0 is small for most substances it can be assumed in the first approximation that $\alpha_e = \pi/2$. For ϵ to be small, it is necessary that $\delta = r_1 - r_2 \ll r_2$. Now (3.192), (3.197) and (3.199) can be rewritten with an allowance for the assumptions made:

$$h = (R + \epsilon) \, \text{arch} \, \frac{r_2 + \delta}{R_2 + \epsilon} \tag{3.205}$$

$$h = R_2 \, \text{arch} \, \frac{r_2}{R_2} \tag{3.206}$$

$$\sqrt{\frac{\beta}{2\epsilon} - \frac{1}{h}} = \eta V. \tag{3.207}$$

Simultaneous solution of (3.205) and (3.206) and employment of the linear terms with respect to ϵ and δ alone gives:

$$\epsilon = \frac{\delta}{q - \sqrt{q^2 - 1}\ \text{arch}q} \tag{3.208}$$

where $q = r_2/R_2$. Note that the denominator in (3.208) becomes zero when $R_2 = R_2^*$. In the vicinity of R_2^* the quadratic terms should be taken into consideration when calculating ϵ and if $R_2 \to R_2^*$, then $\epsilon \to \sqrt{2\delta r_2}\frac{\sqrt{q^{*2}-1}}{q^{*2}}$. However, the formula (3.208) for ϵ will be used here after excluding the narrow range in the vicinity of R_2^* from consideration. Substitution of (3.208) into (3.207) gives:

$$h = \left[\sqrt{\frac{\beta}{2\delta}\left(q - \sqrt{q^2 - 1}\ \text{arch}\ q\right)^{1/2}} - \eta V\right]^{-1} \tag{3.209}$$

By solving (3.206) and (3.209) simultaneously the equilibrium values of h and R_2 can be found (Figure 3.47a). As has already been noted, depending on the parameter values various solutions can exist. The curves described by (3.206) and(3.209) are tangents (the dashed line in Figure 3.47a) under the following condition:

$$\beta r_2 \approx \delta/r_2(1 + r_2\eta V)^{3/2} \tag{3.210}$$

If the point of tangency does not lie in the vicinity of $q = 1$ (this condition can be obtained by equating functions (3.206) and (3.209) to their derivatives). With $\beta r_2 < \delta/r_2(1 + r_2\eta V)^{3/2}$, the whole curve of (3.209) lies above the curve of (3.206) and simultaneous solutions do not exist. If $\beta r_2 > \delta/r_2(1 + r_2\eta V)^{3/2}$, two solutions exist. With increasing β and V decreasing, the root of Type 2 approaches the value of R_2^* and as soon as it reaches this value it disappears (with $R_2 < R_2^*\epsilon < 0$, which has no physical sense). The condition of the second-solution disappearance can be estimated in the following way. With $R_2 \approx R_2^*$, $h \approx r_2$, $\epsilon \approx \sqrt{2\delta r_2}\frac{\sqrt{q^{*2}-1}}{q^{*2}} \approx (\delta r_2)^{1/2}$. Substitution of these values for h and ϵ into (3.207) reveals that with $\beta r_2 < \sqrt{\delta/r_2(1 + r_2\eta V)^2}$, no solutions exist. Thus it is finally obtained that for $\beta r_2 < \delta/r_2(1 + r_2\eta V)^{3/2}$ no roots can exist, for $\delta/r_2(1 + r_2\eta V)^{3/2} < \beta r_2 < \sqrt{\delta/r_2(1 + r_2\eta V)^2}$ two roots exist, for $\beta r_2 > \sqrt{\delta/r_2(1 + r_2\eta V)^2}$, one root exists.

When passing to the stability study in this approximation some singularities of the coefficients A_{ik} should be mentioned and new notations $A_{R_1R_1} \approx A_{R_2R_2} \equiv A_{\epsilon\epsilon}$, $A_{hR_1} = -A_{hR_2} \equiv A_{h\epsilon}$ should be introduced.

Subtraction of the second equation of (3.186) from the first one reveals that $\delta\dot{\epsilon} = A_{\epsilon\epsilon}\delta\epsilon + (A_{R_1h} - A_{R_2h})\delta h$. The third equation of (3.186) has the form $\delta h = A_{h\epsilon}\delta\epsilon + A_{hh}\delta h$. Then for the solutions to be stable it is necessary and sufficient that the following inequalities should be satisfied:

$$A_{R_2R_2} < 0 \tag{3.211}$$

$$A_{\epsilon\epsilon} + A_{hh} < 0 \tag{3.212}$$

$$A_{\epsilon\epsilon}A_{hh} - (A_{R_1h} - A_{R_2h})A_{h\epsilon} > 0. \tag{3.213}$$

Now (3.213) can be rearranged by analogy with (3.201):

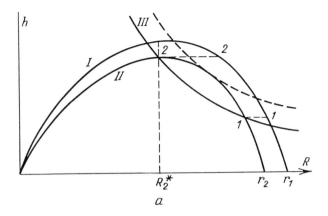

a

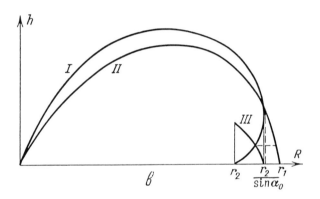

b

Fig. 3.47. Simultaneous solution of the capillary (1, 2) and heat 3 problems: $1 - h(R_1)$ is plotted from (3.197); $2 - h(R_2)$ is plotted from (3.198); $3 - h(R_2)$ is plotted from (3.209); $\alpha_0 = \pi/2$ (a); $\alpha_0 < \pi/2$ (b).

$$A_{\epsilon\epsilon} A_{h\epsilon} \frac{d\epsilon}{dR_2} \left[\left(\frac{dR_2}{dh} \right)_C - \left(\frac{dR_2}{dh} \right)_T \right] > 0 \qquad (3.214)$$

where $\frac{d\epsilon}{dR_2} < 0$ is a derivative of the function $\epsilon(R_2)$ from (3.208), $(\frac{dR_2}{dh})_C$ and $(\frac{dR_2}{dh})_T$ are derivatives of the function $R(h)$ from (3.206) and (3.209), respectively. The root of Type *1* (Figure 3.47a) is stable since inequalities (3.211)–(3.214) are satisfied for it. The root of Type *2* is unstable as (3.214) and hence (3.213) are violated for it. The roots of Types *1* and *2* (Figure 3.47a) correspond to the roots of Types *3* and *4* in the case under consideration (Figure 3.46).

Now let $\alpha_0 \approx \pi/2 - \psi_0 (\psi_0 \ll \pi/2)$. The range of $q < 1$ will be considered. The use of the terms linear with respect to ψ_0 gives the following expression for ϵ:

$$\epsilon = \frac{\delta - 2\psi_0 r_2 - \sqrt{q^2 - 1}/q}{q - \sqrt{q^2 - 1} \text{ arch } q}. \qquad (3.214a)$$

If $\frac{2\psi_0}{\delta/r_0} < 1$ no principle changes as compared with the case of $\psi_0 = 0$ can be found. The solutions change by a small value. Thus, e.g.,

$$h = h_0 \left(1 - \frac{r_2}{\delta} \psi_0 \frac{\sqrt{q^2 - 1}}{q} \right),$$

where h_0 denotes the h-value for $\psi_0 = 0$. If $\frac{2\psi_0}{\delta/r_0} > 1$, such point q_0 exists for which $\epsilon = 0$ (Figure 3.47b). With $q_0 < q^*$, the solutions that make physical sense ($R_1 > R_2$) lie within the range of $q_0 > q > q^*$. In the range of $q < q_0$ only one stable solution exists, it can be found by substituting (3.214a) into (3.207) and by solving this equation simultaneously with (3.206). Analysis of the range of $q < 1$ does not offer new types of solutions. The solutions that lie within the interval of $q^* < q < q^{**}$ are unstable as inequality (3.211) is violated. For $q > q^{**}$, inequalities (3.211)–(3.213) cannot be used to study stability as the condition $A_{R_1 R_1} \approx A_{R_2 R_2}$ under which they were obtained are not satisfied as can be seen from Table 3.2. The general conditions (3.189)–(3.191) should be used for the analysis. However, it should be noted that this range is rather narrow. Indeed, the point q^{**} corresponds to the condition $\frac{\partial h}{\partial \alpha_{02}} = 0$. By differentiating (3.199) $q^{**} = \exp(1/\psi_0)$ is obtained. Thus, consideration of the range of $R_2 < R_2^{**} = r_2 \exp(-1/\psi_0)$ makes sense only for those substances with not particularly small growth angles.

3.16.5. *Conclusions*

When *tubes of large diameters* ($r_1 \gg 1$) are crystallized a stable solution always exists. Apparently, this situation accounts for the relative simplicity of such tube production as compared with the growth of plates, rods and tubes of small diameters.

When *tubes of small diameters* ($r_1 \ll 1$) are crystallized the problem can have two, one or no solutions depending on the values of the process parameters. The crystallization process of tubes of $R_1 > R_1^*$ and $R_2 > R_2^*$ turns out to be stable (Table 3.2). In this case, if two solutions exist one of them is unstable. The solutions for $R_1 > R_1^*$ and $R_2 > R_2^*$ are unstable. Based on this fact, the process of tube closing-up observed during crystallization can be represented in the following way. If R_2 becomes smaller than R_2^* the process loses its stability, which results in further R_2 decrease until the tube closes up and then a rod with $R_1 > R_1^*$ starts to grow, its growth being stable as was shown above. The possibility of stable growth of tubes with $R_2 > R_2^*$ is interesting. Practical growth of such profiles will meet considerable difficulties connected with the limitedness of the range of stability, however, that is wider for substances with considerable growth angles.

3.17. **Influence of Melt Pressure Variations on Crystallization Stability**

In previous investigations on the stability of TPS the crystal dimension R and the crystallization front height h were used as the variable parameters. Another parameter, the pressure d, is added to them and our system becomes a system of three degrees-of-freedom [124]. The linearized set of equations (1.3) relating the relaxation or perturbation-growth rate and the magnitude of these perturbations takes the following form for the case under consideration:

$$\delta \dot{R} = A_{RR}\delta R + A_{Rh}\delta h + A_{Rd}\delta d \qquad (3.215)$$

$$\delta \dot{h} = A_{hR}\delta R + A_{hh}\delta h + A_{hd}\delta d \qquad (3.216)$$

$$\delta \dot{d} = A_{dR}\delta R + A_{dh}\delta h + A_{dd}\delta d \qquad (3.217)$$

Equation (3.215) that is expanded from Equation (1.39) can be shown as follows:

$$\delta \dot{R} = -V \left(\frac{\partial \alpha_0}{\partial R}\delta R + \frac{\partial \alpha_0}{\partial h}\delta h + \frac{\partial \alpha_0}{\partial d}\delta d \right)$$

Hence

$$A_{RR} = -V \frac{\partial \alpha_0}{\partial R}, \quad A_{Rh} = -V \frac{\partial \alpha_0}{\partial h}, \quad A_{Rd} = -V \frac{\partial \alpha_0}{\partial d}.$$

Equation (3.216) is an expanded form of Equation (1.48). Its coefficients are derivatives of the generalized temperature gradient with respect to R, h and d, respectively. The coefficient A_{dd} can be regarded as *zero* here since the temperature gradient in the crystal is explicitly independent of the melt pressure.

The coefficients of Equation (3.217) are found from the mass-balance condition. Here, $A_{dh} = 0$, $A_{dd} = 0$, $A_{dR} = -VC$, where C is the positive constant depending on the ratio of the crystal cross-section area to the melt free-surface area and can be easily calculated from geometric considerations.

The necessary and sufficient condition for the stability of the set of equations (3.215)–(3.217) is that the Routh-Hurvitz conditions having the following form here should be satisfied:

$$A_{hh} + A_{RR} < 0, \qquad (3.218)$$

$$(A_{hh} + A_{RR})(A_{hh}A_{RR} - A_{Rh}A_{hR} - A_{Rd}A_{dR})$$
$$- A_{hh}A_{Rd}A_{dR} > 0, \qquad (3.219)$$

$$A_{hh}A_{Rd}A_{dR} > 0. \qquad (3.220)$$

It follows from the previous analysis, that $A_{Rh} < 0, A_{hR} > 0$. The conditions $A_{hh} < 0, A_{RR} < 0$ can be satisfied by growing a crystal from overheated melt and by creating definite boundary conditions at the shaper free edge. From the form of the coefficient A_{dR} it follows that $A_{dR} < 0$. Thus, it can be assumed that condition (3.218) is satisfied. Then, to satisfy condition (3.220) it is sufficient that A_{Rd} should be positive. It is easy to check that when $A_{Rd} > 0$ inequality (3.219) also holds.

From here it follows that to ensure stability of our system it is necessary that $\partial \alpha_0 / \partial d$ should be negative.

3.17.1. Small Bond Number

The equation of the liquid-meniscus surface in the growth of such crystals takes the form of (1.14). The first integral of this equation is prescribed by (3.103).
The wetting boundary condition on the shaper wall is as follows:

$$z' \big|_{r=R} = -\text{tg}\,\alpha_0, \qquad \alpha_0 = \pi/2 - \psi_0,$$

$$z' \big|_{r=r_0} = -\text{tg}\,\alpha_1, \qquad \alpha_1 = \Theta - \pi/2.$$

Using these boundary conditions, from (3.103) it can be obtained that:

$$R^2 d - R \sin \alpha_0 = r_0^2 d - r_0 \sin \alpha_1.$$

From here it follows that

$$\frac{\partial \alpha_0}{\partial d} = -\frac{1}{\cos \alpha_0} \frac{r_0^2 - R^2}{R} < 0. \tag{3.221}$$

The catching boundary condition at the shaper free edge is as follows:

$$z' \big|_{r=R} = -\text{tg}\,\alpha_0, \qquad z \big|_{r=r_0} = 0.$$

Using (3.103) and these boundary conditions the following expression for the crystallization front height h can be derived:

$$h = z(R) = \int_R^{r_0} \frac{R^2 d - R \sin \alpha_0 - r_0^2 d}{\sqrt{r^2 - (R^2 d - R \sin \alpha_0 - r_0^2 d)^2}} dr. \tag{3.222}$$

The quantity h from (3.222) is a function of α_0 and d, then $\partial \alpha_0 / \partial d = -h'_d / h'_{\alpha_0}$. Differentiation of (3.222) in respect to the parameters d and α_0 gives:

$$h'_{\alpha_0} = \int_R^{r_0} r^2 R \cos \alpha_0 B^{-3/2} dr, \qquad h'_d = \int_R^{r_0} r^2 (r^2 - R^2) B^{-3/2} dr.$$

Here $B = r^2 - (R^2 d - R \sin \alpha_0 - r_0^2 d)^2$. Taking into consideration that $R \leq r \leq r_0$ it is easy to make sure that $h'_{\alpha_0} > 0$, $h'_d > 0$, and hence $\partial \alpha_0 / \partial d < 0$.

3.17.2. Large Bond Number

The equation of the liquid-meniscus surface in the growth of such crystals is specified by (1.13). The first integral of this equation takes the following form:

$$(z - d)^2 + (1 + z'^2)^{-1/2} = C_2. \tag{3.223}$$

Here C_2 is the arbitrary constant of integration.

   The catching boundary condition at the shaper free edge is as follows:

$$z \mid_{r=r_0} = 0, \qquad z' \mid_{z=h} = -\text{tg}\alpha_0.$$

Let the shaper-to-growing crystal clearance be denoted by $l_0 = r_0 - R$. With an allowance for the boundary conditions given above, from (3.223) the following expression can be obtained for l_0:

$$l_0 = \int_0^h [\cos\alpha_0 - (h-d)^2 - (z-d)^2] F^{-1/2} dz. \qquad (3.224)$$

Here

$$F = 1 - [\cos\alpha_0 - (h-d)^2 - (z-d)^2]^2.$$

The quantity l_0 is a function of α_0 and d, therefore $\partial\alpha_0/\partial d = -l'_d/l'_{\alpha_0}$. Differentiation of the integral (3.224) with respect to the corresponding parameters gives:

$$l'_d = 2\int_0^h (h-z) F^{-3/2} dz, \qquad l'_{\alpha_0} = \int_0^h \sin\alpha_0 F^{-3/2} dz$$

Since $h > z$, it is easy to make sure that $l'_d > 0$ and $l'_{\alpha_0} > 0$, therefore $\partial\alpha_0/\partial d < 0$.

   The wetting boundary condition on the shaper wall is as follows:

$$z' \mid_{z=h} = -\text{tg}\alpha_0; \qquad z' \mid_{r=r_0} = -\text{tg}\alpha_1, \qquad \alpha_1 = \Theta - \pi/2.$$

Equation (1.13) will be used, with the origin of coordinates placed on the free surface level $(d = 0)$. Then a change in pressure will correspond to a change in height of the crystallization front h with respect to the free-surface level. It is obvious that $\Delta h = -\Delta d$. Bearing this in mind, the sign of the derivative $\partial\alpha_0/\partial h$ can be defined.

   The first integral of the equation takes the form of (3.15). With an allowance for the boundary conditions, an expression for the shaper wall to growing crystal clearance l can be obtained from the first integral:

$$l_0 = \int_0^h \frac{\xi}{\sqrt{1-\xi^2}} dz + \int_0^{h_0} \frac{\xi}{\sqrt{1-\xi^2}} dz.$$

Here

$$\xi = h^2 + \cos\alpha_0 - z^2, \qquad h_0 = \sqrt{h^2 + \cos\alpha_0 - \cos\alpha_1}.$$

The quantity l_0 is a function of α_0 and h, therefore $\partial\alpha_0/\partial h = -l'_h/l'_{\alpha_0}$. Since $\Delta h = -\Delta d$, $\partial\alpha_0/\partial d = -l'_h/l'_{\alpha_0}$. Differentiation of the expression for $l_0(h, \alpha_0)$ with respect to the corresponding parameters shows that $l'_h > 0$, $l'_{\alpha_0} < 0$ and hence $\partial\alpha_0/\partial d < 0$.

3.17.3. *Conclusions*

Thus, in all the cases considered above $A_{Rd} > 0$, i.e., the crystallization process is stable if its stability is ensured by capillary and heat phenomena. Therefore, all the earlier conclusions regarding the influence of capillary and heat effects on stability remain valid. Thus, pressure variations without compensating the heat or capillary instability reinforce inequalities (3.218)–(3.220), i.e., they make stable processes more stable and unstable processes more unstable.

3.18. **Selection of Shaping Conditions**

A shaper is the main element of the device for melt growth of profile crystals.

The results of the investigations cited above allow a reasonable approach to the solution of the problem of shaping condition selection. First, the problem of shaper material needs to be solved. This material should not react with the melt but should be wetted by it as much as possible. Application of a melt-wettable material for the shaper facilitates relieves the satisfaction of the catching boundary condition at the free edge, which favors the stability of profiled crystal pulling. To ensure the catching condition for a nonwettable shaper it is necessary that positive pressure and small crystal-to-shaper clearance should be used, which presents additional difficulties.

Crystallization process using a nonwettable shaper can also be carried out under stable conditions with the wetting boundary condition satisfied. However, this version requires satisfaction of three conditions: the growth angle should be close to zero, the shaper material should be almost nonwettable by the melt and the crystal-to-shaper wall clearance should be small.

Shaper design is reduced to defining the configuration of shaper free edges and walls with respect to crystal cross-sections, the crystallization front position being fixed. To design a shaper, the melt capillary constant a (it is the unit of measurement of all linear dimensions), the equilibrium crystallization angle ψ_0 and at least approximately the wetting angle Θ should be known.

Let a thin plate $2R$ thick ($R < 0.5$) be grown. It can be approximately assumed that the meniscus shape when growing such a plate will be described by Equation (1.13) for its flat part and by Equation (1.14) for its edge.

The conditions at the edge of the plate being pulled will be considered. To ensure capillary stability the catching boundary condition should be created at the shaper free edge, the condition $\alpha_1 > \Theta - \pi/2$ should be satisfied for this purpose. α_1 is specified by (3.104). With R and α_0 in (3.104) specified, only r_0 and d can vary. It should be remembered that $r_0/R \geq 2$ to ensure capillary stability in this case. If the shaper is absolutely nonwettable by the melt ($\Theta = \pi$), $\alpha_1 > \pi/2$, the profile curve should have a two-valued projection onto the abscissa-axis. From the formulae given in Section 3.6.2, the crystallization front height h can be calculated.

To ensure the horizontal crystallization front such meniscus shaping conditions for the flat part of the tape should be found that provide the same crystallization front height here as that at the edge.

It should be remembered that the value of h is measured off from the shaper level. Once d is known the crystallization front position relative to the melt free surface can be defined. Now based on (3.16) $d - h = \sqrt{A} - \cos \alpha_0$ can be obtained. From here the constant A is defined and for a flat shaper $\alpha_1 : \alpha_1 = \arccos(A - d^2)$ can be calculated. If the calculated value of $\alpha_1 > \Theta - \pi/2$, the catching condition will be satisfied in this part of the tape for the flat shaper; if $\alpha_1 < \Theta - \pi/2$ the wetting condition is obtained. Then capillary stability should be analyzed using Table 3.1.

A simple example will be considered. Let the plate thickness be equal to 0.2 (it should be remembered that where dimensionless quantities are used to measure linear dimensions, the melt capillary constant a is used as a unit of measurement), the growth angle $\psi_0 = 0$ $(\alpha_0 = \pi/2)$, the wetting angle $\Theta = 135°$. The shaper edge will be placed on the level of the melt free surface. Then the relation $R/r_0 > \sin 45° \approx 0.7$ should be satisfied to have the catching condition at the plate edge. Let $R/r_0 = 0.8$, i.e., the curvative radius of the shaper slot is equal to 0.125. Then $\alpha_1 = 54°$. The crystallization front height $h = 0.1$ arch $1.25 = 0.05$. The integration constant in the equation of the flat-part meniscus $A = 0.0025$. From here $\alpha_1 = \arccos A = 89°50'$. From (3.20) the clearance value for the flat part can be calculated. It gives $l_0 \approx 0.015$.

Let a sapphire plate be pulled ($a = 6$ mm); then the plate thickness is equal to 1.2 mm. In this case the shaper sharp edge has the form of a dumb-bell, the slot width is 1.3 mm, the curvature diameter is equal to 1.5 mm. The crystallization front should be held at the height of 0.3 mm above the shaper level.

If the curvative radius exceeds 0.5 the numerical results obtained should be used to find the melt height in this range (Figure 3.31).

3.19. Experimental Tests of Statements on the Theory of Capillary Shaping

3.19.1. Conditions on the Melt-Crystal Interface [125, 126]

From the condition of growth-angle constancy it follows that if $\alpha_0 = \alpha_e = \pi/2 - \psi_0$, a vertical-walled crystal of constant cross-section grows. If $\alpha_0 > \alpha_e$ (the crystallization front is above the equilibrium position) the crystal should contract, if $\alpha_0 < \alpha_e$ (the crystallization front is below the equilibrium position) the crystal should widen. In both cases, the angle $\delta\alpha_0$ of crystal tapering at any moment is equal to $\alpha_0 - \alpha_e$. This model can be proved by pulling thin sapphire rods from a flat capillary tube used as a shaper.

Figure 3.48a shows a photograph of a thin sapphire rod pulled from a capillary tube of the diameter of 0.8 mm. The diameter of the growing crystal is constant if $\alpha_0 = \alpha_e$. Direct measurements give $\psi_0 = (15 \mp 5)°$. Any change in crystallization conditions by varying either the melt temperature or the pulling rate leads to either crystal contraction (Figure 3.48b, d) or to its widening (Figure 3.48c).

If these changes do not violate the stability of the system the crystal lateral surface becomes vertical again when the transition process is over (Figure 3.48b, c). In

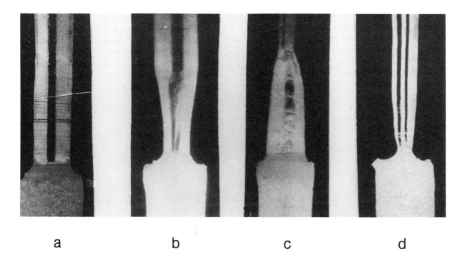

<table>
<tr><td>a</td><td>b</td><td>c</td><td>d</td></tr>
</table>

Fig. 3.48. Melt growth of a thin sapphire filament, the shaper being flat ($x50$): stationary process (a); filament widening (c); filament constriction (b, d).

all cases, the shape of the transition zone corresponds completely to the one shown in Figure 3.41. If in the process of crystal dimension change the growth angle ψ_0 remains constant, the following relation should be valid for the shape of the crystal lateral surface (Figure 1.8):

$$\frac{dR}{dz} = \mathrm{tg}(\alpha_e - \alpha_0) \tag{3.225}$$

Note that differentiation of (3.225) in respect to time can give the law of crystal dimension change with time (1.38).

If the perturbations (the deviation of the radius R from the equilibrium value R^0 and that of the angle α_0 from the equilibrium value α_e) are assumed to be small, then (3.225) takes the form:

$$\frac{dR}{dz} \approx \alpha_e - \alpha_0. \tag{3.226}$$

The quantity α_0 at the point R can be represented in the form of its Taylor series restricting it to the linear terms of the expansion:

$$\alpha_0 = \alpha_e + \frac{\partial \alpha_0}{\partial R}(R - R^0).$$

From here

$$\frac{dR}{dz} = \frac{\partial \alpha_0}{\partial R}(R - R^0). \tag{3.227}$$

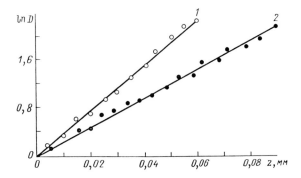

Fig. 3.49.   Sapphire filament constriction (1) and widening (2) in the transient range.

Integration of (3.227) gives the following expression for the transition zone of the crystal:

$$R - R^* = (R^0 - R^*) \exp\left(-\frac{\partial \alpha_0}{\partial R} z\right). \tag{3.228}$$

Here R_0 is the initial value of the crystal radius, R^* is its final value.

Thus, satisfaction of (3.228) can prove the growth angle constancy. Now (3.228) can be rearranged by taking the logarithm of it:

$$\ln D = -\frac{\partial \alpha_0}{\partial R} z. \tag{3.229}$$

Here

$$D = (R - R^*)/(R^0 - R^*).$$

Experimental data are given in Figure 3.49.

An illustrative experiment using a skew capillary tube as a shaper also proves our model of crystal contraction or widening.

By changing the slope between the crystallization front and the surface of the skew shaper by means of the heat conditions, various boundary conditions can be created for various sections of the same crystal. In Figure 3.50a a rod of constant cross-section grows with the capillary tube being skew: the angles ψ_1 on the left and on the right are different ($\psi_1 > \pi/2$ on the left, $\psi_1 < \pi/2$ on the right), but the angles ψ are equal to the growth angle ψ_0; constancy of the diameter is provided by maintaining the angle ψ_0 (or α_e, which is the same) all over the perimeter.

In Figure 3.50b α_0 on the left is equal to the equilibrium value $\alpha_0 = \alpha_e$, on the right $\alpha_0 < \alpha_e$, the right side of the crystal widens; the crystal surface is convex, which corresponds to the range of capillary stability.

In Figure 3.50c α_0 deviation from the equilibrium value on the right side has led to crystal contraction. Due to capillary stability, a rod of constant diameter slightly different from the equilibrium dimension grows.

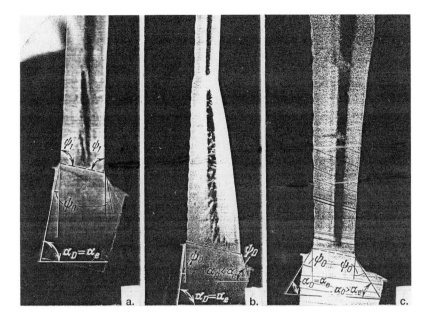

Fig. 3.50. Melt growth of a thin sapphire filament, the shaper being skew filament of constant dimension (a); filament widening from the right (b); filament constriction from the right (c).

The growth angle is approximately equal to $15°$ measured using the photographs and the position of the points corresponding to growth of a crystal of constant diameter.

3.19.2. Profile Curve Heights [125–127]

It is interesting to define whether the liquid column shape qualitatively obeys the laws stated above and to what degree the results of quantitative calculations are applicable in practice.

It was tested when melt-pulling circular rods of high-carbon iron alloy and sapphire. Vacuum of 10^{-4}mm Hg was used. High-carbon iron alloy rods were grown using the "Kristall Ee" installation especially suitable for growing refractory products by TPS. R.f. currents were used for heating. The shaper was made from boron carbonitride. This material is melt steel- and cast iron-resistant and is poorly wetted by them. The diameter of the shaper hole is 11.6 mm, the pressure is equal to zero. The capillary constant is equal to 0.58 cm, and the value of $r_0 = 1$ is obtained for the radius of the shaper hole. The results of comparison of the calculated boundary curves for $\psi_0 = 0$ with the experimental ones are given in Figure 3.51. Since the value γ_{LG} for cast iron strongly depends on the dopant content, the agreement should be considered as good. Moreover, our data can be used to estimate the value of

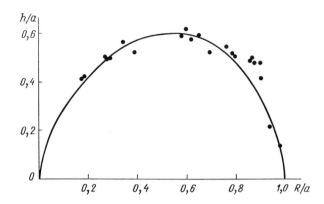

Fig. 3.51. Crystallization front height vs. the diameter of the cylindrical rod pulled from the high-carbon iron melt according to the diagram of Figure I.3c.

surface tension.

The conditions of sapphire crystallization were analyzed while pulling crystalline rods from a capillary tube with the diameter of 0.8 mm according to the diagram shown in Figure I.3e.

Figure 3.52 shows crystallization front height versus crystalline rod diameter. The capillary constant a is used as the unit of measurement. For sapphire $a = 6$ mm. Experimental points are compared with the theoretical curve $h = f(R) \mid_{\alpha_0=75°}$, plotted from formula (3.124). Experiments prove capillary stability presence for the range of the boundary curve where $\frac{dh}{dR} \mid_{\alpha_0=const} < 0$, and its absence for the range where $\frac{dh}{dR} \mid_{\alpha_0=const} > 0$. The shapes of both the contracting or the widening parts of the crystals exactly correspond to those shown in Figure 3.41. The angle of contraction or widening $\delta\alpha_0 \to 0$ with the capillary stability present, $\delta\alpha_0 \to \pi/2$ in case it is absent.

In the range where capillary stability does not exist a rod of constant cross-section cannot be continuously pulled. In practice, a rod of 0.4 mm could not be grown from a capillary-tube shaper with the diameter of 0.8 mm.

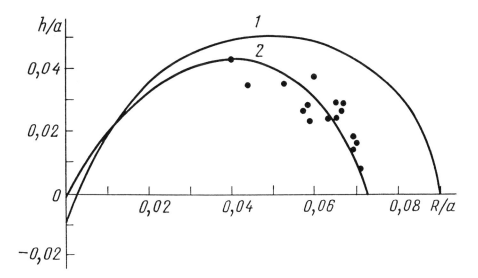

Fig. 3.52. Comparison of the calculated boundary curve $h(R) \mid_{\alpha=75} 0$ and the experimental points for sapphire rods being grown according to Figure I.3e, with the catching boundary condition satisfied at the shaper free edge: 1 – external-edge catching; 2 – internal-edge catching.

Chapter 4

THE VERNEUIL TECHNIQUE

4.1. Growth of Cylinder- and Plate-Shaped Crystals [34]

Stability of cylindrical crystal growth by the Verneuil technique will be analyzed allowing for three degrees-of-freedom: 1) the crystal radius R, 2) the distance between the burner face and the crystallization front l_1 and 3) the liquid-gas interface position l (Figure I.6). Let the melt meniscus height $h = l_1 - l$ measured from the interface to the melt peak point and representing the thickness of the melted layer for flat crystallization fronts be introduced for consideration.

The linearized set of equations (1.3) defined from the Lyapunov criteria of growth process stability takes the following form in the present case:

$$\delta \dot{R} = A_{RR} \delta R + A_{Rl} \delta l + A_{Rh} \delta h \qquad (4.1)$$

$$\delta \dot{l} = A_{lR} \delta R + A_{ll} \delta l + A_{lh} \delta h \qquad (4.2)$$

$$\delta \dot{h} = A_{hR} \delta R + A_{hl} \delta l + A_{hh} \delta h \qquad (4.3)$$

The explicit form of the coefficients of the system should be found.

4.1.1. Determination of the Coefficients A_{RR}, A_{Rl} and A_{Rh}

Comparison of Equations (1.4) and (1.39) gives

$$A_{RR} = -V \frac{\partial \alpha_0}{\partial R}, \quad A_{Rl} = -V \frac{\partial \alpha_0}{\partial l}, \quad A_{Rh} = -V \frac{\partial \alpha_0}{\partial h}, \qquad (4.4)$$

$A_{Rl} = 0$ as the meniscus shape and angle α_0 do not explicitly depend on the liquid-gas interface position.

Functional relation between the angle α_0, the crystal radius R and the meniscus height h is defined from the Laplace capillary equation (1.12).

The specificity of capillary shaping by the Verneuil technique formally reveals itself in the form of the following boundary condition: the line tangent to the melt surface is horizontal at the peak point of the melted layer (Figure 4.1):

167

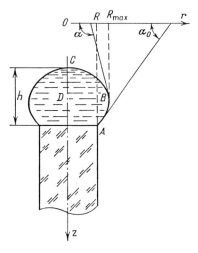

Fig. 4.1. Capillary shaping for the Verneuil technique.

$$dz/dr \mid_{r=0} = 0 \qquad\qquad (4.5)$$

The second boundary condition is specified at the interface and follows from the condition of growth angle constancy.

$$dz/dr \mid_{r=R} = -\mathrm{tg}\alpha_0 \qquad\qquad (4.6)$$

Small Bond number $(R \ll 1)$. Firstly, the case where the crystal radius is considerably smaller than the capillary constant will be discussed. Then the melt weight in Equation (1.12) can be disregarded; it takes the form of (1.14). Profile curve AC by the rotation of which round the z-axis the melted layer shape is obtained will be divided into two portions: AB and BC. Point B is selected in such a way that the line tangent to it could be the vertical. The maximum radius of the melted layer cross-section BD will be denoted by R_m. At point B:

$$dz/dr \mid_{r=R} = \infty. \qquad\qquad (4.7)$$

The function describing the shape of curve AB satisfies the following equation:

$$z_1'' = -r^{-1}(1 + z_1'^2)^{3/2}(\sin\alpha - 2dr) \qquad\qquad (4.8)$$

and for curve BC this equation takes the following form

$$z_2'' = -r^{-1}(1 + z_2'^2)^{3/2}(2dr - \sin\alpha) \qquad\qquad (4.9)$$

The first integrals of Equations (4.8) and (4.9) can be written as follows:

$$-r^2 d - (rz_1')(1 + z_1'^2)^{-1/2} = r\sin\alpha - r^2 d = C_1 \qquad\qquad (4.10)$$

$$r^2 d + (rz_2')(1 + z_2'^2)^{-1/2} = r^2 d - r \sin\alpha = C_2. \tag{4.11}$$

Since $z_1 = z_2$ at point $B, C_1 = C_2$. Besides, $C_1 = C_2 = 0$ as $r = 0$ at point C. Now the melt meniscus height $h = OD + DC$ can be easily defined:

$$h = \int_R^{R_m} r^2 d[r^2 - (r^2 d)^2]^{-1/2} dr - \int_{R_m}^0 r^2 d[r^2 - (r^2 d)^2]^{-1/2} dr$$

$$= d^{-1}[1 - 2(1 - R_m^2 d)^{1/2} + (1 - R^2 d^2)^{1/2}].$$

From (4.7) it follows that $R_m d = 1$ and from (4.6) $d = \sin\alpha_0 R^{-1}$ can be obtained. Taking this into consideration the expression defining $\alpha_0(R, h)$ can be derived:

$$h = R(1 + \cos\alpha_0)(\sin\alpha_0)^{-1}. \tag{4.13}$$

From here taking into account that

$$\partial\alpha_0/\partial h = (\partial h/\partial\alpha_0)^{-1}$$

and

$$\partial\alpha_0/\partial R = (-\partial\alpha_0/\partial h)(\partial h/\partial R)$$

the following equation can be derived:

$$\partial\alpha_0/\partial h = -\sin^2\alpha_0[R(1 + \cos\alpha_0)]^{-1} \tag{4.14}$$

$$\partial\alpha_0/\partial R = -\sin\alpha_0 R^{-1} \tag{4.15}$$

The capillary coefficients have the following forms:

$$A_{RR} = -VR^{-1}\sin\alpha_0 < 0 \tag{4.16}$$

$$A_{Rh} = -VR^{-1}\sin\alpha_0(1 + \cos\alpha_0)^{-1} > 0. \tag{4.17}$$

Large Bond number $(R \geq 1)$. In this case the Laplace equation (1.12) is not solvable by quadrature and numerical calculation is required. However, qualitatively the behavior of the relation can be estimated in the following way. For the meniscus parts where $R \ll 1$ and $R \gg 1$ analytical solutions can be obtained. Then, by joining the functions and their derivatives at the point $R = 1$ a common solution can be obtained. The meniscus height tends to be asymptotic to $(1 + \cos\alpha_0)^{1/2}$ with R increasing. For this approach the greatest deviation from the true relation $h(R)$ is observed in the vicinity of the point $R = 1$, however, it avoids difficulties of the numerical methods and provides qualitative representation of the relation behavior. Figure 4.2 shows qualitative behavior of the relation $h(R)$ for various values. The initial portion of each curve is a straight line defined from Equation (4.13). Within the whole range of crystal dimensions $\partial\alpha_0/\partial R > 0$, $\partial\alpha_0/\partial h < 0$ and correspondingly the signs of the capillary coefficients do not change: $A_{RR} < 0$ and $A_{Rh} > 0$.

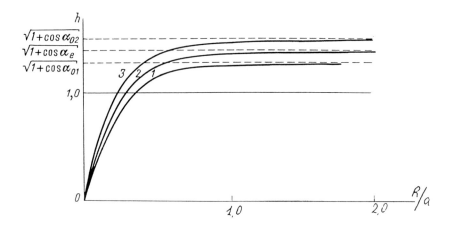

Fig. 4.2.   Melt meniscus height vs. crystal diameter for various angles α_0: $\alpha_{01} > \alpha_e > \alpha_{02}$.

4.1.2. Determination of the Coefficients A_{ll}, A_{lR} and A_{lh}

The mass-balance conditions in the crystal-melt system can be represented in the following form:

$$W_0 = \pi \rho_s R^2 \left(V - \frac{dl_1}{dt} \right) + \pi \rho_L R^2 \left(\frac{dl_1}{dt} \frac{dl}{dt} \right). \tag{4.18}$$

Here W_0 is the charge mass fed into the melted layer per unit time, ρ_s and ρ_L are solid- and liquid-phase densities, respectively. The first component in the right side of (4.18) denotes the substance mass crystallizing per unit time, while the second component is the change in mass of the melted layer. Assumption that $\rho_s = \rho_L = \rho$, made for simplicity, gives:

$$\frac{dl}{dt} = -W(\pi \rho R^2)^{-1} + V. \tag{4.19}$$

Under stationary conditions $dl/dt = 0$ and $V = W_0[\pi\rho(R^0)^2]^{-1}$. When equilibrium is departed from, linearization of (4.19) gives:

$$dl = 2W_0(\pi\rho R^3)^{-1}\delta R. \tag{4.20}$$

Comparison of this equation with (4.2) gives:

$$A_{lR} = 2W_0(\pi\rho R^3)^{-1} > 0, \qquad A_{ll} = 0, \qquad A_{lh} = 0.$$

When calculating the coefficient A_{ll} the charge-mass distribution in the falling flow was assumed to have the form of a delta function (Figure 4.3, curve 1) and with the crystal radius changing the amount of the substance fed to the melt remains unchanged. In a real situation, the charge-mass distribution in the flow along the

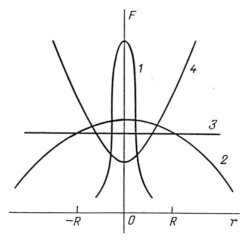

Fig. 4.3.   The function $F(r)$ (charge distribution in the gas flow) depending on the burner design.

muffle cross-section in the crystallization area depends on a number of factors (on the design of the equipment used in particular (Figure 4.3, curves 2–4)) and the substance mass fed into the melt per unit time depends on the crystal radius:

$$W_0 = 2\pi\beta \int_0^R F(r)r\,dr. \tag{4.21}$$

Here $F(r)$ denotes the charge-mass distribution in the flow, β-coefficient of flow rate prescribed by the feeder. Substitution of (4.12) into (4.19) and linearization give:

$$A_{lR} = \beta R^{-3}[2\int_0^R F(r)r\,dr - R^2 F(r)]. \tag{4.22}$$

Possible forms of the function $F(r)$ are depicted in Figure 4.3. It is obvious that if the function $F(r)$ reaches its maximum when $r = 0$ the sign of A_{lR} will not change. If $F(r) = const$ then $F(r) = F(R)$ and $A_{lR} = 0$. Finally, if the minimum of the charge-mass distribution in the flow is observed with $r = 0$, $A_{lR} < 0$. When the charge is fed from the central nozzle of the burner (a widely used practice) the maximum of the charge distribution in the flow lies in the center of the burner muffle (curve 2, Figure 4.3), therefore $A_{lR} > 0$.

4.1.3. Determination of the Coefficients A_{hR}, A_{hl} and A_{hh}

The rate of meniscus height change $\delta\dot{h}$ is specified by Equation (1.48). Liquid- and solid-phase temperature gradients can be defined from Equation (1.40). The crystal is

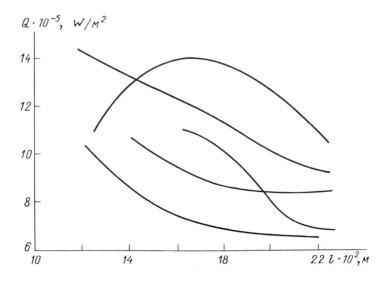

Fig. 4.4.   Heat flow density vs. the distance between the burner and the crystal face for various burner designs [128].

assumed to be cylinder-shaped, its perimeter-to-cross-section area ratio being equal to $F = 2R^{-1}$.

When formulating the heat problem for TPS the liquid-phase temperature on the level of the shaper free edges was specified as the boundary condition. This boundary condition for the Verneuil technique does not correspond to the real situation. Crystal displacement in the furnace muffle results in melt temperature change on its surface. Specificity of the heat conditions of the technique under consideration will be allowed for by specifying the density Q of the heat flow fed from the burner onto the surface of the melted layer. With the gas flow specified, the density of the heat flow depends on the distance between the burner and the level of the melt surface. Figure 4.4 taken from [128] demonstrates the validity of this conclusion. The function $Q(l)$ is determined by the burner design and the gas rate.

Some part of the heat flow fed onto the melt is used for charge melting directly on the melt surface. It is equal to $\eta \mathcal{L} V$, where $\mathcal{L}$ is the melting heat per unit change volume (hereafter to simplify calculations the charge melting heat will be assumed to be equal to the crystal melting heat), η - is the coefficient allowing for the charge state $(0 \le \eta \le 1)$. When $\eta = 0$ the charge is fed into the melt in completely liquid state. When $\eta = 1$ the charge surface starts melting on the melt surface. Thus, the boundary conditions for liquid-phase temperature calculation have the following form:

$$\lambda_L \frac{dT_L^*}{dz}\Big|_{z=1} = -Q + \eta \mathcal{L}\left(V - \frac{dl}{dt}\right), \qquad T_L^*\Big|_{z=l+h} = T_0^*. \qquad (4.23)$$

Here $T_i^* = T_i - T_e$ is the reduced temperature, the index $i = L$ relates to the liquid phase, $i = S$ relates to the solid phase, $T_0^* = T_0 - T_e$ is the reduced crystallization temperature, T_0 denotes the crystallization temperature, T_e is the environmental gas temperature.

Solid-phase temperature distribution is calculated using the following conditions:

$$T_S^* \mid_{z=l+h} = T_0^* \tag{4.25}$$

$$\frac{dT_S^*}{dz} \mid_{z \to \infty} = 0. \tag{4.26}$$

By solving the problem liquid- and solid-phase temperature gradients can be found:

$$-\lambda_L G_L = \left[Q - \mathcal{L} \left(V - \frac{dl}{dt} \right) \right] [1 - \operatorname{th}(\zeta_L h)]$$
$$\times \exp(\zeta_L h) - \lambda_L \zeta_L T_0^* \operatorname{th}(\zeta_L h) \tag{4.27}$$

$$-\lambda_S G_S = \lambda_S \zeta_S T_0^*. \tag{4.28}$$

Here

$$\zeta_i = \sqrt{2\mu_i (\lambda_i R)^{-1}}, \qquad i = L, S. \tag{4.29}$$

One-dimensional analysis is valid for small Bioh numbers, therefore $\zeta_i R \ll 1$. As was shown above (Figure 4.2), $h \approx R$ or $h < R$ and $\zeta_i h \ll 1$. Substitution of the temperature gradients G_L and G_S from (4.27) and (4.28) into (1.48) gives the following expressions for the heat coefficients:

$$A_{hR} = \frac{1}{\mathcal{L}} \left[\lambda_L T_0^* \zeta_L^2 h \frac{1}{R} + \frac{(Q - \eta \mathcal{L} V) \zeta_L h^2}{2R} \right] - A_{lR}(1 - \eta) > 0 \tag{4.30}$$

$$A_{hl} = \frac{1}{\mathcal{L}} \frac{dQ}{dl} \tag{4.31}$$

$$A_{hh} = -\frac{1}{\mathcal{L}} \left[\lambda_L T_0^* \zeta_L^2 + (Q - \eta \mathcal{L} V) \zeta_L^2 h \right] \tag{4.32}$$

When calculating the coefficients (4.30)–(4.32) it was assumed that perturbations of the crystal radius and of the interface position do not practically influence the solid-state temperature gradient as the crystallization front displacement is negligible as compared with the crystal length, and the radius changes only in the vicinity of the crystallization front.

4.1.4. Analysis of the Process-Stability Conditions

For the system of equations to be stable it is necessary and sufficient that the Routh-Gurvitz conditions [108] should be satisfied. For this system they take the following form:

$$A_{RR} + A_{hh} < 0 \tag{4.33}$$

$$A_{Rh} A_{lR} A_{hl} < 0 \tag{4.34}$$

$$-(A_{RR} + A_{hh})(A_{hh} A_{RR} - A_{Rh} A_{hR}) + A_{Rh} A_{lR} A_{hl} > 0 \tag{4.35}$$

Since A_{RR} and A_{hh} are negative, (4.33) holds.

The coefficient A_{Rh} is positive and therefore for (4.34) to hold, the coefficients A_{lR} and A_{hl} should be opposite in sign. As was shown above, when the charge is fed from the central nozzle of the burner $A_{lR} > 0$, then the coefficient A_{hl} should be negative. It means that in the vicinity of the crystallization zone the density of the heat flow fed onto the melt surface should satisfy the condition $\frac{dQ}{dl} < 0$, i.e., the heat-flow density should decrease with the distance from the burner face increasing. For (4.35) to be satisfied the necessary and sufficient conditions are the following:

$$A_{RR} A_{hh} - A_{Rh} A_{hR} > 0 \tag{4.36}$$

$$-(A_{RR} + A_{hh})(A_{hh} A_{RR} + A_{Rh} A_{hR}) > | A_{Rh} A_{lR} A_{hl} | . \tag{4.37}$$

Taking into consideration that $A_{RR} = -\frac{\partial h}{\partial R} A_{Rh}$, where the value of $\frac{\partial h}{\partial R}$ is calculated from the capillary problem (Figure 4.2) the following expression can be derived:

$$A_{RR} A_{hh} - A_{Rh} A_{hR} = \frac{\partial h}{\partial R} \frac{1}{\mathcal{L}} \left[\lambda_L T_0^* \zeta_L^2 + (Q - \eta \mathcal{L} V) \zeta_L^2 h \right]$$

$$\frac{1}{\mathcal{L}} \left[\frac{\lambda_L T_0^* \zeta_L^2 h}{l} + \frac{(Q - \eta \mathcal{L} V) \zeta_L^2 h^2}{2R} \right] + A_{lR}(1 - \eta). \tag{4.38}$$

Let $R \ll 1$, then taking into account that $\partial h / \partial R \approx h / R$ (in accordance with (4.13) h is linearly dependent on R when small radii are used) from (4.38) it can be obtained that $A_{RR} A_{hh} - A_{Rh} A_{hR} > 0$. From here for (4.37) to hold it is sufficient that

$$\left| \frac{dQ}{dl} \right| < \frac{V}{R}. \tag{4.39}$$

It is an approximate evaluation that underestimates $| \frac{dQ}{dl} |$ calculated from (4.37). However, this value of $| \frac{dQ}{dl} |$ deliberately satisfies (4.35) and besides (4.39) makes clear physical sense: for the process to be stable change in the heat flow density Q along the furnace muffle in the vicinity of the growth zone at the distance of an order of R should not exceed the crystallization heat. When this requirement is met and the condition $| \frac{dQ}{dl} | < 0$ is satisfied the process stability is provided for $R \ll 1$.

If $R > 1$, then $\partial h / \partial R$ decreases much faster than h / R (Figure 4.2) does and starting with some values in the range of $R > 1$ crystallization process stability gets lost. From (4.38) it follows that the stability range can be slightly widened by η decreasing, i.e., the charge should be melted before feeding it into the melt. It can be achieved by decreasing the size of powder particles, e.g., by means of its careful separation. Decrease of $| \frac{dQ}{dl} |$ contributes to stability-range widening, however, all these measures do not ensure the process stability when growing crystals of large diameters.

For corundum the capillary constant $a = 6$ mm. Proceeding from the analysis carried out one can state that when corundum crystals are grown crystallization process stability will exist up to the diameters of (10–12) mm. Experience in commercial growth of long thin corundum rods of constant diameters of (3–6) mm for the watch industry is in good agreement with this estimate.

4.1.5. *Plate Growth*

When growing plates whose thickness does not exceed the capillary constant a by the Verneuil technique, the meniscus height h is linearly dependent on the thickness, i.e., it is described by Equation (4.13), where R is the plate half-thickness, which is similar to the case of thin-rod growth. If the heat zone agrees with the plate geometry the heat problem also remains unchanged and hence the conclusion regarding the stability of thin rod crystallization also holds for thin plate crystallization.

4.2. Stability Analysis – Based Automation of the Verneuil Technique [131]

As was shown above, when passing to crystals with diameters exceeding the capillary constant crystallization stability is lost. It means that to grow large crystals of constant cross-sections in situ control of the parameters is required. Usually, an operator controls the parameters by changing charge feed or crystal sinking rate using his experience and intuition.

When developing systems of automatic control of growing-crystal diameters, a problem of the laws of automatic control of the process parameters under some changes in crystal dimensions arises. Up to now the required laws of parameter control were defined from the results of empirical search as the process model has not been developed.

It will be shown that theoretical derivation of the laws of process parameter control ensuring maintenance of the crystal cross-section specified can be performed on the basis of process stability analysis. In this instance the controllable parameters side by side with the crystal radius, the liquid-gas interface position and the meniscus shape can be required as the degrees-of-freedom of the process. For the Verneuil technique the density of the heat flow from the burner Q, the rate of crystal sinking V and the charge flow rate W_0 can be used as controllable parameters. Usually, after the crystal has already widened from the seed size to the desired diameter, control is provided by changing the charge flow rate. The controllable charge flow rate W will be regarded as an additional degree-of-freedom. In this case, the linearized set of equations defining stability of the crystallization process with variable charge flow rate takes the following form:

$$\delta \dot{R} = A_{RR}\delta R + A_{Rl}\delta l + A_{Rh}\delta h + A_{RW}\delta W \qquad (4.40)$$

$$\delta \dot{l} = A_{lR}\delta R + A_{ll}\delta l + A_{lh}\delta h + A_{lW}\delta W \tag{4.41}$$

$$\delta \dot{h} = A_{hR}\delta R + A_{hl}\delta l + A_{hh}\delta h + A_{hW}\delta W \tag{4.42}$$

$$\delta \dot{W} = A_{WR}\delta R + A_{Wl}\delta l + A_{Wh}\delta h + A_{WW}\delta W \tag{4.43}$$

Equation (4.43) is the law of charge flow rate control in its general form. It is quite obvious that in this case the linear law is selected from the variety of possible laws of control. The law of control provides maintenance of the constant crystal cross-section if the set of equations (4.40)–(4.43) is stable. Hence, the problem is to define the coefficients of Equation (4.43) proceeding from the necessity for the set of equations (4.40)–(4.43) to be stable.

The results obtained in the previous paragraph where the Verneuil technique was analyzed for stability will be used to define the coefficients A_{ij}. Let growth of crystals of large diameters ($R \gg 1$) be considered. Then $\partial \alpha_0 / \partial R \to 0$ and $\partial \alpha_0 / \partial h < 0$ and does not depend on the crystal diameter. Therefore, it is assumed that $A_{RR} = 0$, $A_{Rh} > 0$. The coefficients A_{Rl} and A_{RW} are equal to zero as the meniscus shape and the angle α_0 do not depend on the liquid-gas interface position and on the charge flow rate.

The coefficients of Equation (4.41) are defined from the law of mass conservation in the crystal-melt system. The mass-balance equation takes the form of (4.18). Substitution of the variable flow rate W into Equation (4.19) instead of the constant charge flow rate W_0 and linearization give

$$\delta \dot{l} = 2W(\pi\rho R^3)^{-1}\delta R - (\pi\rho R^2)^{-1}\delta W \tag{4.44}$$

From here,

$$A_{lR} = \frac{2W}{\pi\rho R^3} > 0, \quad A_{ll} = 0, \quad A_{lh} = 0, \quad A_{lW} = -\frac{1}{\pi\rho R^2} < 0.$$

In contrast to the case when the charge flow rate is constant, $A_{lW} \neq 0$ here.

The coefficients of Equation (4.42) are the same as the coefficients of Equation (4.3) and their signs are defined above: $A_{hR} > 0$, $A_{hl} < 0$, $A_{hh} < 0$. The coefficient $A_{hW} = 0$, as the meniscus height is not explicitly dependent on the charge flow rate. To distinguish the coefficients included into the law of charge flow rate control from the coefficients of Equation (4.40)–(4.42) the following notations should be introduced: $A_{RW} = K_R$, $A_{lW} = K_l$, $A_{hW} = K_h$, $A_{WW} = K_W$. $K_h = 0$ will be selected since the law of charge flow rate control into which K_h is included is of no practical interest because of the difficulties of monitoring the change in the meniscus height δh.

For the set of Equations (4.40)–(4.43) to be stable, it is necessary and sufficient that the Routh-Gurvitz conditions be satisfied. For a set of equations with four degrees-of-freedom these conditions take the following form:

$$q_1 > 0 \tag{4.45}$$

$$q_1 q_2 - q_3 > 0 \tag{4.46}$$

Stable laws of control	Unstable laws of control
1. $\partial W/\partial t = K_{\dot{l}}\partial l/\partial t + K_w \delta W,$ for $K_l > 0,\ K_w < 0.$	1. $\partial W/\partial t = K_R \delta R + K_{\dot{l}}\partial l/\partial t,$ for K_R and K_l of any sign .
2. $\partial W/\partial t = K_R \delta R + K_{\dot{l}}\partial l/\partial t + K_w \delta W,$ $K_{\dot{l}} > 0,\ K_w < 0,\ K_R < 0$	2. $\partial W/\partial t = K_{\dot{l}}\delta l$.
	3. $\partial W/\partial t = K_R \delta R$.
3. $\partial W/\partial t = K_{\dot{l}}\partial l/\partial t + K_l \delta l,$ $K_l > 0,\quad K_{\dot{l}} < 0$	4. $\partial W/\partial t = K_R \delta R + K_w \delta W$

TABLE. 4.1

$$q_1 q_2 q_3 - q_3^2 - q_1^2 q_4 > 0 \tag{4.47}$$

$$q_4 > 0. \tag{4.48}$$

Here q_1, q_2, q_3, q_4 are the coefficients included in the characteristic equation (1.4) of (1.3). In the present case they are equal to:

$$q_1 = (K_W + A_{hh}) \tag{4.49}$$

$$q_2 = K_W A_{hh} - K_l A_{lW} - A_{Rh} A_{hR} \tag{4.50}$$

$$q_3 = K_W A_{Rh} A_{hR} + K_l A_{hh} A_{lW} - A_{Rh} A_{hl} A_{lR} \tag{4.51}$$

$$q_4 = K_W A_{Rh} A_{hl} A_{lR} + K_l A_{Rh} A_{lW} A_{hR} - K_R A_{Rh} A_{lW} A_{hl} \tag{4.52}$$

The coefficients A_{ij} are known. Where (4.45)–(4.48) are to be satisfied, a set of the coefficients K_j that provide stability of the set of equations (4.40)–(4.43) and hence growth of crystals of constant cross-sections can be obtained (Table 4.1). The laws of charge flow rate control allowing maintenance of the crystal diameter specified are referred to as stable. The law of control $\delta \dot{W} = K_l \delta l + K_{\dot{l}} \delta \dot{l}$ can be obtained if $\delta \dot{l} = A_{lR} \delta R + A_{lW} \delta W$ is taken into account.

When experimentally checking the theoretically derived laws of control, two series of crystals were grown [131]. The first series was grown using the stable law $\delta \dot{W} = K_l \delta l + K_{\dot{l}} \delta \dot{l}$, the second one was grown using the unstable law $\delta \dot{W} = K_l \delta l$. The charge flow rate W is proportional to the voltage of the motor of the vibrating device of the charge feeder: $W = \beta U$, where β is the flow-rate coefficient. Figure 4.5 shows crystals grown using both the stable law of charge flow rate control and the unstable one.

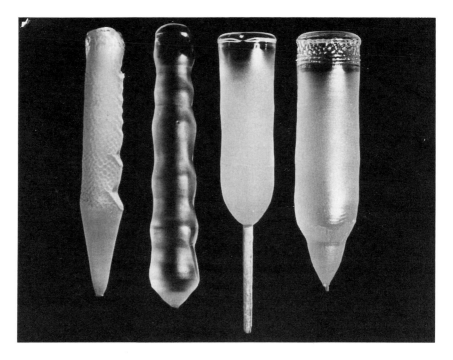

Fig. 4.5.   Corundum crystals grown by stable (to the right) and unstable (to the left) laws of charge control.

4.3. Optimization of the Crystallization Process While Growing Crystals under Unstable Conditions

Optimal growth conditions in the unstable range are such that they provide minimum increment rates for $\delta R, \delta l$ and δh. It enables the specified crystal diameter to be maintained by varying the process parameters more smoothly. This is very important since sharp changes in the process parameters aimed at providing the crystal dimensions specified, e.g., change in the charge flow rate while growing ruby crystals, will lead to local changes in the optical characteristics of the crystal.

For unstable crystallization processes at least one of the roots S_1, S_2, S_3 of the characteristic equation of the set (4.1)–(4.3) has a positive real part. The problem is to define how the process parameters should be varied to provide a decrease of the positive real part of the roots of the characteristic equation. Calculations were made for a corundum crystal with a diameter of 40 mm.

It was stated that the increment rate of the deviations $\delta R, \delta l$, and δh from the stationary quantities R^0, h^0 and l^0, tended to decrease firstly, when the temperature of the burner muffle wall increased; secondly, when the gradient of the heat-flow density from the burner along the furnace muffle decreased with the condition $\frac{dQ}{dl} < 0$ satisfied and thirdly, when the irregular density distribution of the charge flow falling

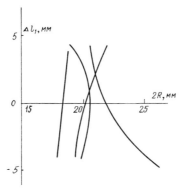

Fig. 4.6. Crystal diameter vs. crystallization front displacement for various growth conditions.

on the melted layer decreased. These requirements of the crystallization conditions are in good agreement with the experimental data regarding the optimal growth conditions [129, 130]. In the experiments described in [129, 130] such hydrogen- and oxygen-flow rate in the channels of a tree-channel burner was set that the condition $\frac{dR}{dl} < 0$ could be satisfied (Figure 4.6). Crystals grown under these conditions exhibited a smoother surface and improved optical and structural characteristics. The condition $\frac{dR}{dl} < 0$ is identical to the condition $\frac{dQ}{dl} < 0$. Indeed, a crystal grown closer to the burner will have a larger diameter only if the heat flow density Q increases when approaching the burner. Decrease of the heat-flow density gradient that, as was shown by the analysis carried out, in turn decreases the increment rate of $\delta R, \delta l$ and δh can in principle be achieved by increasing the oxygen-hydrogen contact perimeter in the flame of a burner, e.g. using a multitube-type burner or by preheating the oxygen and hydrogen before feeding them into the burner. It was stated that preheating the gas before feeding it into the burner and increasing the furnace muffle temperature improved both the structural and optical characteristics of crystals.

4.4. Growth of Tube-Shaped Crystals [35]

4.4.1. *Problem Formulation*

Figure 4.7 shows an idealized diagram of growing tubular crystals shaped by tubular powder flow in the oxygen-hydrogen flame which also has a tubular shape. In contrast to rod-shaped crystal growth, the process of tube growth exhibits not three but four degrees-of-freedom. The outer tube radius R_1, its inner radius R_2, the liquid-gas interface position l and the crystallization front position l_1 can vary independently. The meniscus height $h = l_1 - l$ can be defined as the distance along the vertical from the interface surface to the melted layer peak. As will be shown below, from the

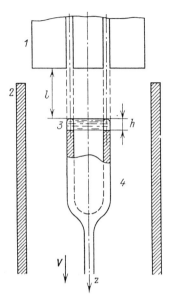

Fig. 4.7.   Growth of one-end-sealed tube-shaped crystals by the Verneuil technique: 1 – burner, 2 – muffle, 3 – melt, 4 – crystal.

solution of the Laplace capillary equation it follows that the height of the meniscus measured outside the tube, h_1, is always larger than the height of the meniscus measured inside the tube, h_2 (Figure 4.8). While analyzing the process stability it will be assumed that the crystallization front displaces parallel to itself when deviations from the stationary growth occur. It means that $\delta h_1 = \delta h_2 = \delta h$, where δh_1 and δh_2 denote changes in height of the outer meniscus h_1 and of the inner one h_2 when the stationary conditions are violated. The linearized set of equations (1.3) describing stability of the crystallization process under consideration takes the following form:

$$\delta \dot{R}_1 = A_{R_1 R_1} \delta R_1 + A_{R_1 R_2} \delta R_2 + A_{R_1 l} \delta l + A_{R_1 h} \delta h \tag{4.57}$$

$$\delta \dot{R}_2 = A_{R_2 R_1} \delta R_1 + A_{R_2 R_2} \delta R_2 + A_{R_2 l} \delta l + A_{R_2 h} \delta h \tag{4.58}$$

$$\delta \dot{l} = A_{l R_1} \delta R_1 + A_{l R_2} \delta R_2 + A_{ll} \delta l + A_{lh} \delta h \tag{4.59}$$

$$\delta \dot{h} = A_{h R_1} \delta R_1 + A_{h R_2} \delta R_2 + A_{hl} \delta l + A_{hh} \delta h. \tag{4.60}$$

4.4.2. Capillary Shaping

Figure 4.8 gives a diagram of capillary shaping of the process under study.
Comparison of Equation (1.39) with (4.57) and (4.58) gives:

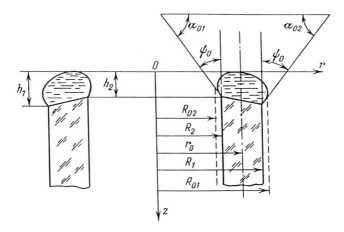

Fig. 4.8. Capillary shaping in tubular crystal growth by the Verneuil technique.

$$A_{R_1R_1} = -V\frac{\partial\alpha_{01}}{\partial R_1} \tag{4.61}$$

$$A_{R_1R_2} = -V\frac{\partial\alpha_{01}}{\partial R_2} \tag{4.62}$$

$$A_{R_1h} = -V\frac{\partial\alpha_{01}}{\partial h} = -V\left(\frac{\partial\alpha_{01}}{\partial h_1} + \frac{\partial\alpha_{01}}{\partial h_2}\right) \tag{4.63}$$

$$A_{R_2R_1} = -V\frac{\partial\alpha_{02}}{\partial R_2} \tag{4.64}$$

$$A_{R_2R_2} = -V\frac{\partial\alpha_{02}}{\partial R_2} \tag{4.65}$$

$$A_{R_2h} = V\frac{\partial\alpha_{02}}{\partial h} = V\left(\frac{\partial\alpha_{02}}{\partial h_1} + \frac{\partial\alpha_{02}}{\partial h_2}\right) \tag{4.66}$$

$$A_{R_1l} = 0 \tag{4.67}$$

$$A_{R_2l} = 0 \tag{4.68}$$

The coefficients A_{R_1l} and A_{R_2l} are equal to *zero* since the meniscus shape and the angles α_{01} and α_{02} are not explicitly dependent on the liquid-gas interface position.

Functional relations between the angles α_{01} and α_{02}, the tube radii and meniscus height can be defined from the Laplace capillary equations. The boundary conditions are specified at the melted layer peak: $dz/dr\ |_{r=r_0} = 0$, where r_0 is the projection of the melted layer peak on the r-axis; on the interface line outside the tube: $dz/dr\ |_{r=R_1} = -\mathrm{tg}\alpha_{01}$, and inside the tube: $dz/dr\ |_{r=R_2} = \mathrm{tg}\alpha_{02}$. If a tube with an outer radius smaller than the capillary constant is considered, the shape of the meniscus profile can be defined from the Laplace axially-symmetric equation from which the melt column weight is excluded (1.14). Solution of this equation with the boundary conditions mentioned above gives the following expressing for h_1 and h_2:

$$h_1 = \int_{R_{01}}^{R_1} (C - r^2 d)[r^2 - (C - r^2 d)^2]^{-1/2} dr$$

$$+ \int_{R_{01}}^{r_0} (C - r^2 d)[r^2 - (C - r^2 d)^2]^{-1/2} dr \qquad (4.69)$$

$$h_2 = \int_{R_{02}}^{r_0} (C - r^2 d)[r^2 - (C - r^2 d)^2]^{-1/2} dr$$

$$+ \int_{R_{02}}^{R_2} (C - r^2 d)[r^2 - (C - r^2 d)^2]^{-1/2} dr \qquad (4.70)$$

where C is the integration constant, d denotes the pressure in the melt dependent on the meniscus curvature alone and R_{01} and R_{02} are the portions cut by the vertical tangents to the meniscus on the r-axis (fig.4.8). All these quantities and r_0 are expressed in terms of R_1 and R_2, α_{01}, α_{02} and using the first integral and simple geometrical relations in the following way:

$$d = (R_1 \sin \alpha_{01} + R_2 \sin \alpha_{02})(R_1^2 - R_2^2)^{-1} \qquad (4.71)$$

$$C = R_1 R_2 (R_1 \sin \alpha_{01} + R_2 \sin \alpha_{02})(R_1^2 - R_2^2)^{-1} \qquad (4.72)$$

$$r_0 = \{R_1 R_2 (R_1 \sin \alpha_{02} + R_2 \sin \alpha_{01})(R_1 \sin \alpha_{01} + R_2 \sin \alpha_{02})^{-1}\}^{1/2} \qquad (4.73)$$

$$R_{01} = \{2(R_1 \sin \alpha_{01} + R_2 \sin \alpha_{02})\}^{-1} \{(R_1^2 - R_2^2) + [(R_1^2 - R_2^2)^2$$

$$+ 4 R_1 R_2 (R_1 \sin \alpha_{02} + R_2 \sin \alpha_{01})(R_1 \sin \alpha_{01} + R_2 \sin \alpha_{02})]^{1/2}\} \qquad (4.74)$$

$$R_{02} = \{2(R_1 \sin \alpha_{01} + R_2 \sin \alpha_{02})\}^{-1} \{(R_2^2 - R_1^2) + [(R_2^2 - R_1^2)^2$$

$$+ 4 R_1 R_2 (R_1 \sin \alpha_{02} + R_2 \sin \alpha_{01})(R_1 \sin \alpha_{01} + R_2 \sin \alpha_{02})]^{1/2}\} \qquad (4.75)$$

The outer meniscus height h_1, the inner meniscus height h_2 and the partial derivatives defining the capillary coefficients were numerically calculated for various values of R_1, R_2, α_{01}, α_{02}. Table 4.2 gives the calculated results for the specified outer tube radius $R_1 = 0.95$ and fixed angles $\alpha_{01} = \alpha_{02} = 75°$ depending on the change in the tube inner radius R_2 for corundum ($\psi_0 = 13° \mp 2°$) that is the main material grown by this technique. The capillary constant is used as a unit of measurement of linear dimensions.

From the results given in Table 4.2 and the expressions for the capillary coefficients (4.61)–(4.68) it follows that $A_{R_1 R_1} < 0$, $A_{R_1 R_2} > 0$, $A_{R_1 h} > 0$, $A_{R_2 h} < 0$, $A_{R_2 R_1} > 0$. The sign of the coefficient $A_{R_2 R_2}$ depends on the value of R_2. Increase

	$R_1 = 0,95$				$\alpha_{01} = \alpha_{02} = 75\,^\circ$			
R_2	h_1	h_2	$\dfrac{\partial \alpha_{01}}{\partial R_1}$	$\dfrac{\partial \alpha_{01}}{\partial R_2}$	$\dfrac{\partial \alpha_{01}}{\partial h}$	$\dfrac{\partial \alpha_{02}}{\partial R_1}$	$\dfrac{\partial \alpha_{02}}{\partial R_2}$	$\dfrac{\partial \alpha_{02}}{\partial h}$
0,1	1,14	0,15	1,44	−3,43	−0,05	0,89	15,87	16,36
0,2	0,89	0,32	1,64	−2,83	−0,42	1,03	4,22	−8,52
0,3	0,69	0,25	1,89	−2,70	−0,80	1,24	1,02	−6,40
0,4	0,53	0,24	2,23	−2,79	−1,26	1,53	−0,22	−5,68
0,5	0,40	0,22	2,67	−3,06	−1,89	1,94	−1,21	−5,62
0,6	0,29	0,19	3,34	−3,58	−2,85	2,58	−2,18	−6,14
0,7	0,18	0,14	4,49	−4,60	−4,58	3,70	−3,51	−7,26
0,8	0,10	0,09	7,07	−7,04	−8,62	6,25	−6,19	−11,36
0,9	0,03	0,03	19,40	−18,92	−29,45	−18,45	−19,40	−31,57

TABLE. 4.2

of the tube inner radius causes monotonous decrease of the coefficient $A_{R_2 R_2}$ from positive to negative values. For other values of R_1, R_2, α_{01} and α_{02} the values of the derivatives change, however, their signs and behaviors depending on R_2 are the same.

4.4.3. Heat and Mass Transfer. Determination of the Coefficients of Equation (4.59)

The rate of liquid-gas interface displacement is defined from the mass balance condition in the crystal-melt system in the same way as for rod-shaped crystals. The assumption that the densities of solid and liquid phases are equal gives:

$$\frac{dl}{dt} = -W_0[\pi\rho(R_1^2 - R_2^2)]^{-1} + V \tag{4.76}$$

Linearization of (4.76) gives:

$$\delta \dot{l} = 2W_0 R_1(\pi\rho)^{-1}(R_1^2 - R_2^2)^{-2}\delta R_1$$
$$-2W_0 R_2(\pi\rho)^{-1}(R_1^2 - R_2^2)^{-2}\delta R_2. \tag{4.77}$$

From comparison of (4.59) and (4.77) follows:

$$A_{lR_1} = 2W_0 R_1/\pi\rho(R_1^2 - R_2^2) > 0$$
$$A_{lR_2} = -2W_0 R_2/\pi\rho(R_1^2 - R_2^2) < 0.$$

Determination of the coefficients of Equation (4.60). In the given case the heat coefficients are defined in the same way as for cylindrical crystals. The boundary conditions and a number of assumptions made while determining the heat coefficients in the cylindrical crystal stability analysis are the same. However, if for a rod $F = 2/R$ in the heat-conductivity equation (1.40), for a tube under the condition of no heat removal from the internal surface $F = 2R_1/(R_1^2 - R_2^2)$. Then it is obtained that

$$A_{hR_1} = \mathcal{L}^{-1}\left[0.5(Q - V)\zeta_L h^2 \left(-\frac{d\zeta_L}{dR_1}\right) + 2\lambda_L T_0^* \zeta_L h \left(-\frac{d\zeta_L}{dR_1}\right)\right] \quad (4.78)$$

$$A_{hR_2} = \mathcal{L}^{-1}\left[0.5(Q - V)\zeta_L h^2 \left(-\frac{d\zeta_L}{dR_2}\right) + 2\lambda_L T_0^* \zeta_L h \left(-\frac{d\zeta_L}{dR_2}\right)\right] \quad (4.79)$$

$$A_{hl} = \mathcal{L}^{-1}\frac{dQ}{dl} \quad (4.80)$$

$$A_{hh} = -\mathcal{L}^{-1}\left[(Q - V)\zeta_L^2 h + \lambda_L T_0^* \zeta_L\right] \quad (4.81)$$

Here $\zeta_L = [2\mu R_1(R_1^2 - R_2^2)^{-1}\lambda_L^{-1}]^{1/2}$ and mean height $h = 0.5(h_1 + h_2)$ is regarded as the meniscus height, the sign of A_{hl} is defined by the sign of dQ/dl. It is assumed that $dQ/dl < 0$. It means that the heat-flow density increases while approaching the burner as l is measured from the burner.

Two limiting cases will be considered. Let $R_2 \ll R_1$, then

$$\zeta_L \approx [2\mu/\lambda_L(R_1 - R_2)]^{1/2}, \quad \frac{d\zeta_L}{dR_1} = -0.5\zeta_L R_1^{-1}, \quad \frac{d\zeta_L}{dR_2} = -0.5\zeta_L R_2^{-1}.$$

The coefficients A_{hR_1} and A_{hR_2} assume the following forms:

$$A_{hR_1} = \mathcal{L}^{-1}\left[0.5(Q - V)\zeta_L^2 h^2 R_1^{-1} + \lambda_L T_0^* \zeta_L h R_1^{-1}\right] > 0 \quad (4.82)$$

$$A_{hR_2} = -\mathcal{L}^{-1}\left[0.5(Q - V)\zeta_L^2 h^2 R_2^{-1} + \lambda_L T_0^* \zeta_L h R_2^{-1}\right] < 0. \quad (4.83)$$

If $R_2 \ll R_1$, then $\zeta \approx \{2\mu(\lambda_L R_1)^{-1}\}^{-1/2}$, and $d\zeta_L/dR_2 = 0$. Hence, in this case $A_{hR_1} > 0$, while $A_{hR_2} = 0$.

4.4.4. *Process-Stability Analysis*

As was shown above, for solution of a set with four unknowns to be stable it is necessary and sufficient that the conditions (4.45)–(4.48) be satisfied. In the case under consideration the coefficients of the characteristic equation of the set of Equations (4.57)–(4.60) are equal to:

$$q_1 = -(A_{R_1 R_1} + A_{R_2 R_2} + A_{hh}) \quad (4.84)$$

$$q_2 = A_{R_1 R_1} A_{R_2 R_2} + A_{R_1 R_1} A_{hh} + A_{R_2 R_2} A_{hh} - A_{R_1 R_2} A_{R_2 R_1}$$

$$- A_{R_2 h} A_{hR_2} - A_{R_1 h} A_{hR_1} \quad (4.85)$$

$$q_3 = A_{R_1 R_1} A_{R_2 h} A_{hR_2} + A_{hh} A_{R_1 R_2} A_{R_2 R_1}$$

$$-A_{R_1R_1}A_{R_2R_2}A_{hh} + A_{R_2h}A_{lR_2}A_{hl}$$

$$+A_{R_1R_2}A_{R_2h}A_{hR_1} - A_{R_2R_1}A_{R_1h}A_{hR_2}$$

$$-A_{R_1h}A_{lR_1}A_{hl} + A_{R_2R_2}A_{R_1h}A_{hR_1} \qquad (4.86)$$

$$q_4 = (A_{R_1R_1}A_{R_2h} - A_{R_2R_1}A_{R_1h})A_{lR_2}A_{hl}$$

$$+(A_{R_2R_2}A_{R_1h} - A_{R_1R_2}A_{R_2h})A_{lR_1}A_{hl}. \qquad (4.87)$$

Stability of the set of Equations (4.57)–(4.60) will be analyzed for a thin-walled tube, i.e., for $R_2 \leq R_1$. In this approximation $A_{lR_1} \approx -A_{lR_2}$ and $A_{hR_1} \approx -A_{hR_2}$. To simplify further analysis the heat coefficients will be expressed in terms of A_{hh}; it is obtained that $A_{hR_1} = -hR_1^{-1}A_{hh}$ and $A_{hR_2} = hR_2^{-1}A_{hh}$. According to (4.81)–(4.83) these are crude estimates overestimating the absolute values of A_{hR_1} and A_{hR_2}. However, such approximation is permissible as where overestimated values of A_{hR_1} and A_{hR_2} satisfy (4.45)–(4.48), smaller magnitudes of these coefficients will deliberately satisfy the inequalities.

The condition (4.47) imposes constraints on the magnitude of the coefficient $\mid A_{hl} \mid = \mid \mathcal{L}^{-1}\frac{dQ}{dl} \mid$. Some critical value of $\mid \frac{dQ}{dl} \mid^*$ exists above which (4.47) does not hold, i.e., it is required that the change in the density of the heat flow along the furnace muffle should not be too sharp. Where $0 < \mid \frac{dQ}{dl} \mid < \mid \frac{dQ}{dl} \mid^*$, analysis of (4.45)–(4.48) using the numerical values of the capillary coefficients given in Table 4.2 shows that for $R_2 \leq R_1$ all the inequalities hold and hence the solution of the set of Equations (4.57)–(4.60) is stable. This means that the crystallization process is stable.

The second limiting case will be considered $(R_2 \ll R_1)$. As can be seen from Table 4.2, the coefficient $A_{R_2R_2}$ reverses its sign, becomes positive, $A_{lR_2} \rightarrow 0$, $A_{hR_2} \rightarrow 0$. As follows from (4.87), (4.48) is not satisfied. The first term of (4.87) vanishes, the second one is negative since $A_{hl} < 0$. If just one of the inequalities (4.45)–(4.48) does not hold, the solution of the set of equations (4.57)–(4.60) is unstable. Then the crystallization process during thick-walled tube growth is also unstable.

Proceeding from the results of the previous paragraph where it was shown in particular that growth of a plate less than one capillary constant thick is stable, it is possible to state that crystallization of thin-walled tubes of a large diameter will be stable since a thin-walled tube of a large diameter can be regarded as a plate of finite but small curvature.

Stability of growing thin-walled tubes of small and large outer diameters leads to the following qualitative conclusion. Crystallization of tubes of arbitrary outer diameters is stable if the tube wall thickness is smaller than some critical thickness. This thickness is smaller than the capillary constant and depends both on the heat conditions of the process and on the outer radius of the tube. This crystal thickness of the tube wall increases with the outer diameter increase. The crystallization stability is lost when the tube-wall thickness is larger than the capillary constant.

4.4.5. *Experiments on Growing Single-Crystal Corundum Tubes under Stable Conditions*

In the previous paragraph it was stated that stable growth is possible when the tube wall thickness is smaller than the capillary constant. For corrundum the capillary constant is equal to 6 mm.

In our experiments [35], tubes with an outer diameter of 17 to 25 mm and with walls 3 to 4 mm thick were grown. A crystallization apparatus fitted with a four-channel burner providing charge supply via the central and the periphery channels was used. Crystal growth was initiated from a seed 3 to 4 mm in diameter. It was experimentally stated that the optimal gas distribution in the burner channels is as follows: oxygen-hydrogen-oxygen-hydrogen. Firstly, a seed cone was grown. The cone was widened by feeding the charge through the central channel, the periphery oxygen flow rate being increased. The base diameter of the seed core reached 20-22 mm. As soon as the crystal diameter reached the specified value the charge was fed through the periphery channel. 10–15 min later charge supply from the central tank was cut off and within 30–60 min the rate of the central oxygen flow was reduced. The latter provides smooth transition from a solid crystal to a tube. The sink rate was gradually increased and the process of stationary growth continued. Tubes up to 120 mm long were grown. Usually no parameter control to maintain constant cross-section of the tube was required, i.e., stable growth conditions (the existence of which had been theoretically predicted) could be attained. Figure I.7 shows some of the tubes grown.

Chapter 5

THE FLOATING-ZONE TECHNIQUE

5.1. Problem Formulation

The results of our analysis of stability of crystallization by the Czochralski, TPS and Verneuil techniques cited in the present monograph are original with the exception of a few particular points. The floating-zone technique was first analyzed for crystallization stability in [132]. However, this analysis was not complete as the heat conditions of crystallization were not taken into account and the capillary part of the problem was greatly simplified, i.e, only capillary stability was studied while thermal fluctuations are considered to produce no effect on the process stability. Furthermore, only plate – and cylindrical crystal – growth were investigated from the point of view of capillary phenomenon analysis. The simplest version for a plate was considered, i.e, meniscus under nearly microgravity conditions, while investigation of capillary shaping of meniscus with cylindrical symmetry is based on the results of numerical calculation of liquid zones for one particular case when both the growing crystal and the melting one are of the same size. Capillary stability proved to exist in the cases under consideration and this conclusion was extended to all other versions of the floating-zone technique. It will now be shown that capillary stability does not exist for some crystal dimension and pressures in meniscus [133, 134].

An idealized diagram of crystallization by the floating-zone technique is shown in Figure I.8 (upward pulling). At present the versions of downward pulling are widely used. Various kinds of heating devices are used for melted-zone formation: radiant, induction, electron, and laser heating. Irrespective of the heater design and geometry, a region of maximum heating exists in the middle part of the zone that is a convenient physical point to start measurements when analyzing heat phenomena of this crystallization technique (Figure I.8). It should be remembered that the shaper free edge is used as such a starting point in TPS, and the level of the melt free surface is used as the starting point in the Czochralski technique. In contrast to capillary-fed crystallization (TPS), here the pressure d in the meniscus is determined by melted-zone curvature and is unambiguously related to the zone volume W. In view of this, it is convenient to introduce the melt volume W as one of the degrees-of-freedom when

187

analyzing stability of crystallization by the floating-zone technique, its variation in time being easy to find on the basis of the condition of mass balance. In this case, both W and d can be found by simultaneously solving the capillary and the heat problems with an allowance for the condition of mass balance.

Besides the crystallization front position h_c, the melting front position h_m can vary independently in the floating-zone technique; variation h_m in time can be found on the basis of the heat balance condition at the melting boundary. Overall zone length $h = h_c - h_m$. In this equality the fact that h_m is negative in the coordinate system of Figure I.8. is accounted for h_m by the minus sign.

Thus, in the floating-zone technique four quantities can vary independently, they are the crystal dimension R, the crystallization front position h_c, the melting front position h_m and the zone volume W. The set of equations defining crystallization stability in a general case takes the following form:

$$\delta \dot{R} = A_{RR}\delta R + A_{Rh_c}\delta h_c + A_{Rh_m}\delta h_m + A_{RW}\delta W$$

$$\delta \dot{h}_c = A_{h_c R}\delta R + A_{h_c h_c}\delta h_c + A_{h_c h_m}\delta h_m + A_{h_c W}\delta W$$

$$\delta \dot{h}_m = A_{h_m R}\delta R + A_{h_m h_c}\delta h_c + A_{h_m h_m}\delta h_m + A_{h_m W}\delta W$$

$$\delta \dot{W} = A_{WR}\delta R + A_{Wh_c}\delta h_c + A_{Wh_m}\delta h_m + A_{WW}\delta W \qquad (5.1)$$

5.2. The Heat Conditions

The heat conditions for a crystal being grown and the zone part between the origin of coordinates coinciding with the point of maximum heating and the crystallization front are completely identical to those discussed above for the Czochralski technique. Hence, the coefficients $A_{h_c R}$ and $A_{h_c h_c}$ coincide with the corresponding heat coefficients A_{hR} and A_{hh}. The coefficient $A_{h_c h_m} = 0$ as the temperature gradients in the zone part under consideration do not depend on the melting front position h_m. The rate of melting front displacement can be defined from the heat balance condition at the melting boundary given below:

$$\lambda_L G_{Lm} - \lambda_S G_{Sm} = \mathcal{L}\left(V_m - \frac{dh_m}{dt}\right) \qquad (5.2)$$

Here, V_m is the rate at which the crystal being melted moves. The heat flow from the liquid phase to the melting front should provide melting of the crystal fed with the rate V_m. This leads to a higher liquid-phase temperature gradient at the melting front than the corresponding gradient at the crystallization front. Since the temperature difference $T_m - T_0$ relative to the point of maximum heating is the same in both cases it appears that $h_m < h_c$, this inequality being enhanced with the crystallization rate V_c increasing and correspondingly with the melting rate V_m increasing. For small h_m-values heat removal from the lateral surface of this meniscus part can be neglected in the first approximation and the liquid- phase gradient at the melting front assumes the following form: $G_{Lm} = -(T_m - T_0)/h_m$. Substitution of G_{Lm} into (5.2) gives for $\delta \dot{h}_m$:

$$\delta \dot{h}_m = \frac{\lambda_L (T_m - T_0) \delta h_m}{h_m^2}$$

i.e.,

$$A_{h_m h_m} = \frac{\lambda_L}{\mathcal{L}} \frac{T_m - T_0}{h_m^2}$$

Here, $h_m/h_c \ll 1$ serves as an applicability criterion of the approximation mentioned for which $A_{h_m R}$, $A_{h_m h_c}$ and $A_{h_m W}$ can be considered as equal to *zero*.

Usually, the true meniscus shape is not taken into consideration when heat phenomena are analyzed, it is replaced with a cylinder of the radius R. As regards the floating-zone technique, an independent change of W results in a change in the mean radius of the meniscus and in a corresponding change in the liquid-phase temperature gradient. The sign of the coefficient $A_{h_c W}$ coincides with the sign of $A_{h_c R}$. Indeed, both W and R increasing leads to increases of the heat flow from the liquid-phase towards the crystallization front and to corresponding decrease of the crystallization rate. The influence of the effect mentioned on crystallization stability will be discussed below and in the first approximation it will be assumed that $A_{h_c W} = 0$.

5.3. Mass Balance

The rate of melted zone volume changes can be determined on the basis of the heat balance condition

$$\delta \dot{W} = \pi \rho_S \rho_L^{-1} [r_0^2 (V_m - \dot{h}_m) - R^2 (V_C - \dot{h}_c)] \tag{5.3}$$

Here ρ_S and ρ_L are the densities of the solid and the liquid phases, respectively. Under the stationary-growth conditions $\delta \dot{W} = 0$ and $r_0^2 V_m = R^2 V_C$. From the radius of the crystal being melted r_0, V_m and V_C, the dimension of the stationary growing crystal R^0 can be calculated. When equilibrium is violated $\delta \dot{W} \neq 0$, $\delta R \neq 0$, $\delta h_m \neq 0$, $\delta h_c \neq 0$. Application of (5.1) and (5.3) for the corresponding coefficients gives:

$$A_{WR} = B + C A_{h_c R}, \quad A_{W h_c} = C A_{h_c h_c}, \quad A_{W h_m} = -C r_0^2 R^{-2} A_{h_m h_m},$$

$$A_{WW} = C A_{h_c W}, \quad B = -2\pi \rho_S \rho_L^{-1} RV, \quad C = \pi \rho_S \rho_L^{-1} R^2.$$

It should be stressed that $A_{h_c W}$ is assumed to be equal to zero.

5.4. The Capillary Coefficients

As is shown in [36], the boundary condition of meniscus catching on the edge of the crystal being melted is satisfied at the melting front. Meniscus slope towards the lateral surface of this crystal can be arbitrary in contrast to the growth angle constancy during crystallization.

For these boundary conditions, i.e., angle fixation at the crystallization front, catching at the melting front, the behavior of the capillary coefficients as a function of the relation between the dimensions of the crystal being grown and of the crystal being melted was investigated in [133, 134]. Figure 5.1. shows the summarized results. Axially symmetric menisci of large radii of their cross-section curvature will be taken as examples for our analysis. The crystal radius R as a function of the parameters h, and d for upward crystal pulling was derived for identical menisci when TPS was analyzed:

$$R(h, \alpha_0, d) = r_0 - \int_0^h [C_1 - (z - d)^2] \tag{5.4}$$

where $C_1 = \cos(\alpha_0) + (h - d)^2$ and $F = 1 - [C_1 - (z - d)^2]$. This relation was shown in Figure 3.15. An identical relation for downward crystal pulling is shown in Figure 5.2. (curve breaks are associated with the impossibility for stationary menisci to exist for the growth angles ψ_0 specified).

It should be mentioned that there exists a definite range of stationary meniscus existence ($\alpha_0 = \alpha_e = \pi/2 - \psi_0$) specified by the inequality $F \geq 0$. The maximum achievable meniscus height is defined as $h_{max} = |d| + (1 + \cos \alpha_0)^{1/2}$ for downward crystal pulling where $d = -\sqrt{2}$, and as $h_{max} = |d| + (1 - \cos \alpha_0)^{1/2}$ for upward crystal pulling where $d = \sqrt{2}$. From here, it follows that the maximum achievable stationary- meniscus height is slightly higher for downward crystal pulling. The capillary coefficients can be found:

$$A_{RR} = -V\partial\alpha_0/\partial R, \quad A_{Rh_c} = -V\partial\alpha_0/\partial h,$$

$$A_{Rh_m} = -V\partial\alpha_0/\partial h, \quad A_{RW} = -V\partial\alpha_0/\partial W.$$

Exact differentials of dh and dw can be written in the following way:

$$dh = \left(h'_R\right)_{\alpha_0,d} dR + \left(h'_{\alpha_0}\right)_{R,d} d\alpha_0 + \left(h'_d\right)_{\alpha_0,d} d(d),$$

$$dW = \left(W'_R\right)_{\alpha_0,d} dR + \left(W'_{\alpha_0}\right)_{R,d} d\alpha_0 + \left(W'_d\right)_{\alpha_0,d} d(d). \tag{5.5}$$

Elimination of the common parameter d gives the desired derivatives:

$$(\alpha_0)_W = \frac{W'_d}{R'_{\alpha_0} W'_d - R'_d W'_{\alpha_0}} = \frac{W'_d}{\Delta}, \tag{5.6}$$

$$(\alpha_0)'_W = R'_d \Delta^{-1}, \quad \text{where } \Delta = R'_{\alpha_0} W'_d - R'_d W'_{\alpha_0} \tag{5.7}$$

$$(\alpha_0)'_h = \left(R'_{\alpha_0} W'_d - R'_d W'_{\alpha_0}\right) \Delta^{-1}. \tag{5.8}$$

Taking into account that the zone volume, e.g., for the case of upward pulling is equal to:

$$W(h, \alpha_0, d) = \pi R^2 h + 2\pi R \int\limits_0^h z F_1 dz,$$

where the following notation is introduced:

$$F_1 = [C - (z - d)^2]\{1 - [C_1 - (z - d)^2]\}^{-1/2},$$

partial derivatives included into (5.6)–(5.8) can be found:

$$R'_{\alpha_0} = \int\limits_0^h (\sin \alpha_0) F^{-3/2} dz \qquad (5.9)$$

$$R'_d = 2 \int\limits_0^h (h - z) F^{-3/2} dz \qquad (5.10)$$

$$R'_h = - \operatorname{ctg} \alpha_0 + 2 \int\limits_0^h (d - z) F^{-3/2} dz \qquad (5.11)$$

$$W'_{\alpha_0} = 2\pi R h R'_{\alpha_0} + 2\pi R'_{\alpha_0} \int\limits_0^h z F_1 dz - 2\pi R \int\limits_0^h z(\sin \alpha_0) F^{-3/2} dz \qquad (5.12)$$

$$W'_d = 2\pi R h R'_d + 2\pi R'_d \int\limits_0^h z F_1 dz - 4\pi R \int\limits_0^h z(h - z) F^{-3/2} dz \qquad (5.13)$$

$$W'_h = 2\pi R h R'_h + \pi R^2 + 2\pi R h \operatorname{ctg}(\alpha_0) + 2\pi R'_h \int\limits_0^h z F_1 dz$$

$$-4\pi R \int\limits_0^h z(d - z) F^{-3/2} dz. \qquad (5.14)$$

It is not difficult to show that the denominator of (5.6)–(5.8) is always positive, $h > z$ in all cases, therefore the signs of the derivatives are as follows: $(\alpha_0)'_R > 0$ and $(\alpha_0)'_W < 0$. Numerical calculations are necessary to analyze $(\alpha)'_h$ as the sign of the numerator in (5.8) can vary. Calculations revealed that in all the ranges of meniscus existence for various values of α_0 the derivative $(\alpha)'_h$ is positive in the case of upward crystal pulling and can alter its sign in the case of downward pulling (Figure 5.1).

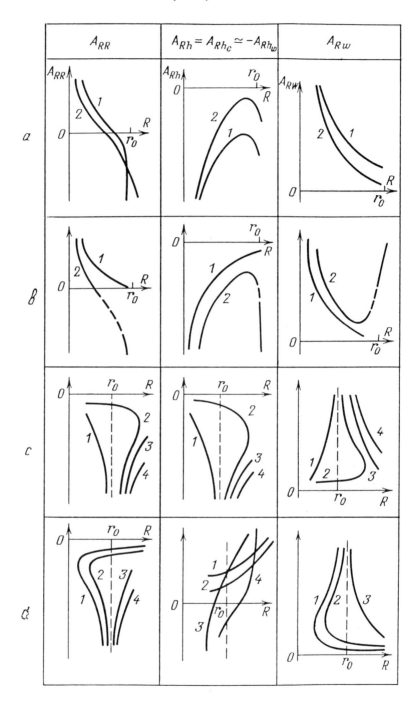

Fig. 5.1. The coefficients A_{ik} vs. R ($\phi_0 = 12°$) for meniscus of various shapes (a–d) under various pressures (1–4) (the value of $r_0 d$ increases with the curve number): one-valued (a) and two-valued (b) meniscus for small Bond numbers; crystals are pulled upwards (c), downwards (d) for large Bond numbers.

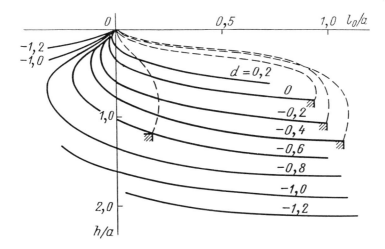

Fig. 5.2. Boundary curves and some profile curves (the dashed lines) for meniscus with large Bond numbers, $\psi_0 = 12°$, under various pressures d.

5.5. Crystallization Stability

With the results of the analysis of the coefficients A_{ik} carried out above taken into consideration, according to the Routh-Gurvitz criterion for the set of equations (5.1) to be stable it is necessary and sufficient that the following inequalities hold.

$$A_{h_m h_m} < 0 \tag{5.15}$$

$$A_{RR} + A_{h_c h_c} < 0 \tag{5.16}$$

$$BA_{RW} + A_{h_c h_c} > 0 \tag{5.17}$$

$$BA_{RR}A_{RW} - A_{RR}A_{h_c h_c}(A_{RR} + A_{h_c h_c}$$
$$+A_{h_c R}(A_{RR} + A_{h_c h_c})(A_{Rh_c} + CA_{RW}). \tag{5.18}$$

If in the inequalities written above the heat coefficients $A_{h_c R}$ and $A_{h_c h_c}$ are formally assumed to be equal to zero, the stability criteria $(\alpha_0)'_R > 0$ and $(\alpha_0)'_W < 0$ coinciding with those from [132] can be obtained. The conditions (5.15)–(5.18) are more general. If $T_m > T_0$ (the zone is overheated) the coefficients $A_{h_m h_m}$ and $A_{h_c h_c}$ are negative, (5.15) and (5.16) always hold, and only under such conditions is stable growth by the floating-zone possible. As was shown above, for all the parameters of the zone $A_{RW} > 0$ and since $B < 0$, (5.17) always holds. Now (5.18) will be analyzed in detail. Using Figure 5.1. one can ensure that the first two terms of this inequality are positive. Inequality (5.18) will be satisfied if in the third term $| A_{Rh_c} | > CA_{RW}$. Using (5.9)–(5.14) the following equations can be derived:

$$CA_{RW} = \pi R^2 V R'_d / \Delta, \tag{5.19}$$

$$A_{Rh_c} = -\frac{V}{\Delta}\left\{ \pi R^2 R'_d + 4\pi R\,\text{ctg}(\alpha_0)\int\limits_0^h (h-z)F^{-3/2}dz + 8\pi R(h-d) \right.$$

$$\left. \times \left[\int\limits_0^h z^2 F^{-3/2}dz \cdot \int\limits_0^h F^{-3/2}dz - \left(\int\limits_0^h z F^{-3/2}dz\right)^2 \right] \right\}. \qquad (5.20)$$

Inequality (5.18) will hold if the second and third terms in (5.20) add up to a positive number. It can be shown that this sum can be negative, i.e., the condition of stable growth (5.18) can be violated for meniscus corresponding to the portions of the boundary curves (Figure 3.15, 5.2.) with positive slopes towards the vertical, i.e., $R'_h > 0$.

Indeed, if $h > d$ the sum is positive as the difference in the square brackets satisfies the Koshi-Bunyakovskiy inequality. In this case, according to (5.11) $R'_h < 0$. Thus, the system is stable outside the interval of $R'_h > 0$ on the boundary curves.

Now let $h < d$ but $R'_d < 0$ as before, i.e., $\text{ctg}(\alpha_0) > 2(d-h)\int_0^h F^{-3/2}dz$. In (5.20) $\text{ctg}(\alpha_0)$ will be replaced with a smaller value equal to the previously written integral. If the sum mentioned in (5.20) is written down in such a way the following condition is derived: $2(d-h)(hI)_1^2 - I_2^2) > 0$ where $I_1 = \int_0^h F^{-3/2}dz$ and $I_2 = \int_0^h z F^{-3/2}dz$. Again (5.18) is satisfied outside the range of the boundary curves where $R'_h > 0$.

It has already been mentioned that in the case of downward crystal pulling the co-efficient A_{Rh_c} varies. When passing to positive A_{Rh_c}-values (5.18) can be violated. In Figure 5.2. the boundary corresponding to sign reversal of A_{Rh_c} can be easily defined. Note that sign reversal of A_{Rh_c} alone cannot cause loss of overall stability.

By comparing the versions of upward and downward crystallization by the floating-zone technique it can be found that for the upward crystallization version a larger portion of the boundary curve (and then the majority of possible crystal dimensions) corresponds to the range of stable growth but in this case the dimension of the crystal being pulled is smaller than that of the crystal being melted and in contrast for the downward crystallization version a larger portion of the boundary curve corresponds to the range of possible instability $R'_h > 0$. Nevertheless, in practice, the downward version is used more frequently when growing crystals by the floating-zone technique since only in this case can crystals of large diameters be pulled under stable conditions using corresponding selection of the diameter of the crystal being melted and the zone volume.

From the point of view of stability, for the floating-zone technique menisci of large curvature are identical to those of TPS in the sense that A_{RR} reverses its sign. In this case, some stability limit exists and it is such that crystals of dimensions smaller than the critical ones cannot grow in a stable way. However, it can be strictly shown that for the floating-zone technique the limit of capillary stability $((\alpha_0)'_R = 0)$ displaces towards the range of smaller crystal dimensions as compared with TPS.

As has already been mentioned, changes in the zone volume W with R and h fixed cause changes in the liquid-phase temperature gradient. Up to now, the influence of this effect on crystallization stability has not been analyzed. However, it can easily be noticed that the effect decreases stability. Indeed, as was shown above, an increase in the zone volume leads to an increase in the heat flow towards the crystallization front and to a decrease of the crystallization rate, which in its turn leads to further increase of W.

Absence of such an effect in TPS growth is therefore certainly an advantage.

Chapter 6

RADIAL INSTABILITY OF WHISKERS

The approach to the stability investigation of growing crystal shape developed in the present monograph can be used to explain causes of radial instability and to describe the shape of initial parts of whiskers grown by the vapor-melt-crystal mechanism. However, in contrast to the previous section, kinetic effects that influence the growth angle value will now play an important part. In the previous sections, these effects were not taken into consideration.

Periodic radial instability of whisker growth (Figure 6.1.) is described and analyzed in detail in [135].

So far there have been two attempts to explain the causes of self-excited oscillations when growing whiskers by the vapor-liquid-crystal mechanism. In [135] radial instability is qualitatively attributed to variations of crystallization front roughness associated with various concentrations of the main crystallizing substance in a melt drop at the whisker top and periodic changes of the 'contact angle' are also considered to be caused by it.

In [136] this explanation is criticized and the variations of the 'contact angle' are considered to be caused by the influence of uncontrollable surfactants available in the gas phase. But the authors ignore small changes in the concentration of the main crystallizing component in the melt.

Here, it will be shown that radial instability can be explained ignoring possible availability of surfactants as a natural consequence of the Gibbs-Thomson dimensional effects when whisker diameters reduce to submicron dimensions [137]. A real experimental silicon-gold system investigated most thoroughly is analyzed with an allowance for its specific features using chloride precipitation process in the $SiCl_4$+ H_2 gaseous mixture in a flowing - gas reactor [135].

The problem of the initial whisker part shaping during crystallization is directly connected with defining the law that relates the slope of the growing-crystal lateral surface and the crystallization conditions. The profile curve of this section can easily be calculated in an isotropic approximation of the free energy of the whisker lateral surface [137].

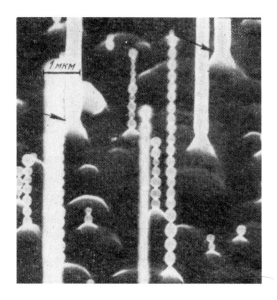

Fig. 6.1.   Silicon whiskers [165].

6.1. Problem Formulation

If the whisker radius R and the main component concentration C in the solution-melt system are selected as independent variables a linearized set of differential equations (1.3) describing the rate of change in the concentration and in the whisker radius for small deviations from the stationary state will formally assume the following form:

$$\delta \dot{R} = A_{RR}\delta R + A_{RC}\delta C, \qquad \delta \dot{C} = A_{CR}\delta R + A_{CC}\delta C. \tag{6.1}$$

The Routh-Gurvitz inequalities for this system take the following form:

$$A_{RR} + A_{CC} < 0, \qquad A_{RR}A_{CC} + A_{RC}A_{CR} > 0. \tag{6.2}$$

The values of the coefficients A_{ij} depend on the selected stationary-state parameters of a real system (whisker diameter, gas-mixture composition and pressure, temperature, etc.). If (6.2) holds, the stationary state relates to the stable-focus type, where the sign of the inequality for the diagonal coefficients is reversed ($A_{RR} + A_{CC} > 0$) while the sign of the inequality $A_{RR}A_{CC} + A_{RC}A_{CR} > 0$ remains unchanged, the dynamic system loses its stability and passes to the unstable focus-type state (Figure 2.1.). It means that limiting cycles corresponding to self-excited oscillations of a real dynamic system can appear on the phase diagram [135]. The next problem of the analysis is to concretize (6.1) and to investigate the conditions satisfying (6.2) [137].

6.2. Whisker Growth

Ignoring the adsorbed-atom surface flows along the lateral surface towards the crystallization front it will be assumed that stationary whisker growth is provided by means of the flow of the main substance fed from the gas phase through the liquid-drop surface towards the whisker top and removed from the drop in the process of crystallization. The concentration C can be defined as the ratio of the main-substance atom number n to the drop volume W, $C = nW^{-1}$. From the condition of impurity mass M constancy in a melt drop, the drop volume can be defined from the following formula:

$$W = M[\rho(1 - \Omega C)]^{-1} \tag{6.3}$$

where ρ is the metal-solvent density, Ω is the volume per solid- or liquid-phase atom. Differentiation of the concentration $\dot{C} = d(n/W)/dt$ with respect to time gives the mass-balance equation:

$$\dot{C} = \frac{1}{2W}\left(JS - \frac{\pi R^2 V}{\Omega}\right). \tag{6.4}$$

Here J is the density of the substance flow from the gas phase, S denotes the melt-drop lateral surface area, V is the crystallization rate of a whisker of the radius R. The factor 1/2 appears in (6.4) as an allowance for specific features of a concrete silicon-gold constitutional diagram, i.e., at the temperature T of crystallization by the vapor-liquid-crystal mechanism equal to $T \simeq 1000^0C$ silicon concentration in gold $C \simeq 50\%$ at. [34].

The crystallization rate V, the dependence of which on the process parameters is not known beforehand but can be calculated from experimental data, is included in (6.4). In [34] it is shown that whisker growth rate is determined by crystallization front kinetics and for a number of substances including silicon whiskers is well described by the quadratic dependence on the difference in chemical potential between gas (μ_G)- and solid (μ_S)0-phases

$$V = b(\mu_G - \mu_S)^2 = b(\Delta\mu_{GS})^2 \tag{6.5}$$

where b is kinetic coefficient.

It was also experimentally determined that initiation of self-excited oscillations is correlated with a bend on the plot of crystallization rate versus whisker diameter, which might be associated with the Gibbs-Thomson dimensional effect [83]. With an allowance for the dimensional effect, the chemical-potential difference in (6.5) can be written as:

$$\Delta\mu_{GS} = kT \ln \frac{P}{P_S^0(T)} - \frac{\Omega\gamma_{SG}}{R} \tag{6.6}$$

where P is the specified gas-phase partial pressure of the main component, $P_S^0(T)$ is the equilibrium pressure at the uncurved crystal surface at the process temperature T, γ_{SG} is the solid-phase free surface energy, k denotes the Boltzmann's constant.

Since crystallization in the silicon-gold system is carried out at low gas-phase oversaturation [135] the substance-flow density through the surface of a melt drop of the radius R can be defined in the following way:

$$J = \beta \left(\Delta \mu^0_{GL} - \frac{2\Omega \gamma_{LG}}{R} \right). \tag{6.7}$$

Here β is the kinetic coefficient, $\Delta \mu^0_{GL}$ denotes the difference in chemical potential between gaseous and uncurved liquid phases, γ_{LG} is the melt free surface energy and allowance will be made for the relation $\Delta \mu^0_{GL} = \Delta \mu^0_{GS} - \Delta \mu^0_{LS}$ or in a different way:

$$kT \ln \frac{P}{P^0_L(T, C)} = kT \ln \frac{P}{P^0_S(T)} - \frac{q \Delta T}{T} \tag{6.8}$$

where $P^0_L(T, C)$ denotes the equilibrium vapor pressure at the uncurved surface of the melt of the composition C at the temperature T, q is the melting heat per atom, ΔT relates to undercooling at the whisker crystallization front. Now from (6.7) and (6.8) an explicit relation between the substance flow density J and the melt concentration C can be defined since the undercooling temperature ΔT is connected with the melt oversaturation by the relation $\Delta T = \eta(C - C_0)$ where η denotes the liquidus-line slope on the constitutional diagram, $C_0(T)$ is the equilibrium concentration of the main component in the melt at the process temperature T.

To derive a closed set of equations describing stationary crystallization of whiskers it is necessary to find the law that relates the slope of the growing-crystal lateral surface and the crystallization conditions. This can be done on the basis of the results obtained in 1.3, 1.4.

In the vicinity of the three-phase line the crystallization front and the crystal lateral surface are severely curved (Figure 1.7a, d). Surface curvatures are the result of the Gibbs-Thomson effect. It can be assumed that on the three-phase line itself $\mu_S = \mu_L$, far from it the lateral surface curvature decreases, i.e., a chemical potential gradient (grad μ_S) associated with the curvature gradient exists. Under the action of this gradient along the crystal surface mass transfer occurs and under stationary conditions the three-phase line and the curved front and crystal lateral surface sections contiguous with it displace as a whole with the rate V.

The curvature of this crystal lateral surface in the vicinity of the three-phase line can be specified by the angle $\chi(s)$, where s is the distance along the surface measured off the three-phase line (Figure 1.7d). To calculate the profile curve of the whisker lateral surface the equation relating the substance flow density J (the number of atoms crossing a unit three-phase line length per unit time) and (grad μ_S) should be used (1.27)

$$J = -\lambda \frac{d\mu_S}{ds} \tag{6.9}$$

where λ is the coefficient of surface mass transfer. In a general case the chemical potential gradient (grad μ_S) comprises the gradients of azimuthal and meridian curvatures. However, it should be noted that the azimuthal curvature $\xi_1(s)$ changes negligibly over a small curved section $\simeq 0.05\mu$ m long [32] and can be estimated from the formula $\xi_1(s) = \cos \chi(s)/R \simeq 1/R$. At the same time the radius of the meridian curvature $\xi_2^{-1}(s)$ over a curved section of the same length ranges from a few micrometers to infinity (Figure 1.7d). It means that in (6.9) the meridian curvature gradient $\xi_2(s) = -d\chi(s)/ds$ is of particular importance. The growth rate along the normal towards the outer surface $V_S = V_C \sin \chi \simeq V_C \chi(s)$ is determined by the substance flow density J (1.28):

$$NV_S = -\frac{d\mu_S}{ds} \tag{6.10}$$

where $N = 1/\Omega$ denotes the number of atoms per unit volume. The equation for chemical potentials of curved liquid and solid phases takes the following form:

$$\mu_L = \mu_L^0 + \frac{2\Omega\gamma_{SG}}{R} \tag{6.11}$$

$$\mu_S = \mu_L^0 - \frac{q\Delta T}{T} + \frac{\gamma_{SG}}{N}(\xi_1 + \xi_2) \tag{6.12}$$

where μ_L^0 is the chemical potential of the uncurved liquid phase. After corresponding substitutions and with an allowance for (1.31) the solution of (6.10) can be represented in the following way:

$$\chi(s) = \chi_0 \exp\left[-\left(\frac{N^2 V_C}{\lambda\gamma_{SG}}\right)^{1/3}\right]. \tag{6.13}$$

Here χ_0 is the angle of the maximum crystal-surface slope towards the direction of whisker growth V_C. Calculation of the equilibrium meridian curvature ξ_2 from the condition $\mu_S = \mu_L$ and Equations (6.10), (6.12):

$$\xi_2 \mid_{s=0} = -\frac{d\chi}{ds}\bigg|_{s=0} = \frac{Nq\Delta T}{T\gamma_{SG}} + \frac{2\gamma_{LG}}{R\gamma_{SG}} - \frac{1}{R}$$

allows easy calculation of the angle $\chi_0 = \chi \mid_{s=0}$ by simple differentiation of (6.13):

$$\chi_0 = k_\psi V_C^{-1/3}\Delta T + k_1 V_C^{-1/3} R^{-1}\frac{\gamma_{LG}}{\gamma_{SG}} - 0.5k_1 V_C^{-1/3} R^{-1} \tag{6.14}$$

where

$$k_\psi = \frac{q}{T}\left(\frac{N\lambda}{\gamma_{SG}^2}\right)^{1/3}, \quad k_1 = 2\left(\frac{\lambda\gamma_{SG}}{N^2}\right)^{1/3}$$

are *constants* for the crystal and solvent materials.

In (6.14) the first term is identical to that of (1.32) and in this case the contribution of the concentrational undercooling $\Delta T = \eta\Delta C$ at the crystallization front to whisker lateral surface curving is taken into consideration, the second and the third terms appear as a result of the allowance for liquid- and solid-phase surface curvatures.

In Figure 1.7d it can be seen that the final lateral surface of the crystal is shaped at the following growth-angle values:

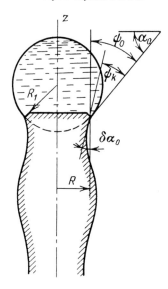

Fig. 6.2.   Shaping of the lateral surfaces of whiskers grown by the VLC mechanism.

$$\psi_k = \psi_0 - \chi_0. \tag{6.15}$$

The equilibrium angle ψ_0 determined by the values of the free surface energy according to (1.23) is included into (6.15). The angle of the final crystal-surface slope towards the growth axis $\delta\alpha_0$ is specified by the slope of the line tangent to the meniscus ψ and the growth angle ψ_k (Figure 6.2) in the following way:

$$\delta\alpha = \psi - \psi_k$$
$$= \psi - \psi_0 + k_\psi V^{-1/3}\eta\Delta C - k_1 V^{-1/3}R^{-1}\left(\frac{\gamma_{LG}}{\gamma_{SG}}\cos\psi - 0.5\right). \tag{6.16}$$

Generally speaking, the direction of the vector of the whisker crystallization rate V does not coincide with the direction of the vertical growth rate V_C, however within the small perturbation analysis being carried out it can be assumed that $V = V_C/\cos(\delta\alpha_0) \simeq V_C$. Thus, the rate of changes in the whisker diameter will take the following form:

$$\delta\dot{R} = V_C\mathrm{tg}(\delta\alpha_0) \simeq V\delta\alpha_0. \tag{6.17}$$

The equations given above are sufficient to describe stationary crystallization of whiskers and to analyze crystallization stability with respect to small perturbations.

6.3. Stability Study

With an allowance for (6.4) and (6.8), the set of differential equations (6.1) can be written in the explicit forms:

$$\delta \dot{R} = V(\delta \alpha_0)'_R \delta R + V(\delta \alpha_0)'_C \delta C \tag{6.18}$$

$$\delta \dot{C} = \frac{1}{2W} \left(J'_R S + J S'_R - \frac{2\pi RV}{\Omega} - \frac{\pi R^2}{\Omega} V'_R \right) \delta R$$

$$+ \frac{JS}{2W} \left(\frac{J'_C}{J} + \frac{S'_C}{S} \right) \delta C. \tag{6.19}$$

From Figure 6.2 melt drop parameters can be calculated. The lateral-surface area is equal to

$$S = 4\pi R_1^2 - 2\pi R_1 h = 2\pi R^2 (1 - \sin \psi)^{-1}. \tag{6.20}$$

The drop volume equal to

$$W = \frac{4}{3} \pi R_1^3 - \frac{1}{6} \pi h (3R^2 + h^2) =$$

$$\frac{1}{3} \pi R^3 [2 + \sin \psi (2 + \cos^2 \psi)] \cos^{-3} \psi. \tag{6.21}$$

Since $S'_C = S'_\psi \psi'_C$, by introducing the function

$$f = \frac{M}{\rho(1 - \Omega C)} - \frac{\pi R^3 [2 + \sin \psi (2 + \cos^2 \psi)]}{3 \cos^2 \psi} = 0$$

the following derivatives can be found:

$$\psi'_C = -\frac{f'_C}{f'_\psi} = \frac{M\Omega}{\rho(1 - \Omega C)^2} \frac{(1 - \sin \psi)^2}{\pi R^3} \tag{6.22}$$

$$S'_C = \frac{M\Omega}{\rho(1 - \Omega C)^2} \frac{2 \cos \psi}{R} = 2\Omega W \frac{\cos \psi}{R}. \tag{6.23}$$

Differentiation of (1.23) gives

$$(\psi_0)'_C = \left(\cos\psi_0 - \frac{1}{\sin \psi_0} \frac{\gamma_{LG}}{\gamma_{LG}} \right) \frac{1}{\gamma_{LG}} (\gamma_{LG})'_C. \tag{6.24}$$

Now the coefficients $A_{ij}(R, C)$ of (6.1) with an allowance for (6.18) and (6.19) in a general form can be written as follows:

$$\frac{1}{V} A_{RR} = \psi'_R - \frac{1}{3} k_\psi V^{-1/3} \eta \Delta C V'_R$$

$$- \frac{1}{3} k_1 V^{-2/3} R^{-1} \left(\frac{\gamma_{LG}}{\gamma_{SG}} \cos \psi - 0.5 \right) V'_R$$

$$- k_1 V^{-1/3} R^{-2} \left(\frac{\gamma_{LG}}{\gamma_{SG}} \cos \psi - 0.5 \right) - k_1 V^{-1/3} R^{-1} \sin \psi \psi'_R \tag{6.25}$$

$$\frac{1}{V}A_{RC} = \psi'_C - (\psi_0)'_C + k_\psi V^{-1/3}\eta + k_1 V^{-1/3} R^{-1}\frac{\cos\psi}{\gamma_{SG}}(\gamma_{LG})'_C -$$

$$-k_1 V^{-1/3} R^{-1}\frac{\gamma_{LG}}{\gamma_{SG}}\sin\psi\psi'_C \tag{6.26}$$

$$A_{CC} = \frac{\beta S}{2W}\left[\frac{2W\gamma_{LG}}{R}\sin\psi\psi'_C - \frac{2W}{R}\cos\psi(\gamma_{LG})'_C - \frac{q\eta}{T}\right]$$

$$+\frac{J\Omega\cos\psi}{R} \tag{6.27}$$

$$A_{CR} = \frac{\beta S}{W}\frac{\Omega\gamma_{LG}}{R^2}\frac{(1-\sin\psi)^2}{\cos\psi} + \frac{1}{2W}\left(JS'_R - \frac{2\pi RV}{\Omega} - \pi\beta\gamma_{SG}\right). \tag{6.28}$$

The behavior of thin and thick whiskers for which the dimensional effects can be disregarded in (6.18) and (6.19) will be compared.

For thick whiskers it is not difficult to estimate the signs of the coefficients $A_{ij}(R, C)$ and then to check satisfiability of the Routh-Gurvitz conditions (6.2):

$$\frac{1}{V}A_{RR} = \psi'_R < 0 \tag{6.29}$$

$$\frac{1}{V}A_{RC} = \psi'_C - (\psi_0)'_C + k_\psi V^{-1/3}\eta > 0 \tag{6.30}$$

$$\frac{W}{JS}A_{CC} = \frac{J'_C}{J} + \frac{S'_C}{S} < 0 \tag{6.31}$$

$$WA_{CR} = JS'_R - \frac{2\pi RV}{\Omega} < 0. \tag{6.32}$$

The signs of the coefficients A_{RR} and A_{CR} in (6.29) and (6.32) are obvious since the corresponding partial derivatives are derived for constant concentration values, which are equivalent to the condition of melt-drop volume constancy. Positivity of the coefficient A_{RC} in (6.30) follows from the following three conditions: $(\gamma_{LG})'_C < 0$, $\eta_0 > 0$ and $\cos\psi_0 > \gamma_{LG}/\gamma_{SG}$ that hold for the silicon-gold system in particular. To find the sign of the second diagonal coefficient, numerical estimate of the terms of the right side of (6.31) is required since the derivatives $S'_C > 0$ and $J'_C < 0$ are of opposite signs. Using (6.20) and (6.21) and taking into account that $\Omega \simeq 10^{-23}\text{cm}^3$, $S'_C/S = \Omega\cos\psi(1-\sin\psi)W/\pi R^3 \simeq 10^{-23}\text{cm}^3$ can be found. To estimate $J'_C/J = -q\eta/T\Delta\mu^0_{LG}$, $\Delta\mu^0_{LG}$ given by (6.8) will be substituted in the fraction denominator by a larger quantity $\Delta\mu^0_{SG}$, so that J'_C/J when estimated should be smaller than the real value. It is obtained that $J'_C/J = -q\eta/kT^2\ln(P/P^0_S) \approx$ -600 cm^{-3} as for silicon $q/kT \sim 4$, $\eta/T \sim 15$ m^3 and $\ln(P/P^0_S) \approx 10^{-1}$ [25].

Thus, from the analysis carried out it follows that for systems of Si- Au-type when thick whiskers are crystalline the natural equilibrium state is that of the stable focus type since the necessary and sufficient Routh-Gurvitz conditions are satisfied.

When analyzing the influence of the dimensional effects on whisker crystallization by the vapor-liquid-crystal mechanism, the corrections in (6.18) and (6.20) resulting from the Gibbs-Thomson effect should be taken into consideration. Thus, when the conditions $\gamma_{LG}/\gamma_{SG} \cos \psi \simeq \cos \psi_0 \cos \psi \leq 0.5$ hold, all the corrections to A_{RR} are positive and increase with whisker dimension decreasing as a function of $\sim R^{-2}$, which is evident from (6.25). It is obvious that for some critical whisker diameter the sign off A_{RR} is positive. Similarly, from (6.27) it follows that dimensional corrections to A_{CC} are also positive and increase as R^{-2}. The coefficient A_{RC} with the decreasing of whisker diameter becomes negative, because the corrections according to (6.26) are negative and increase as $\sim R^{-1}$. The coefficient A_{CR} also reverses its sign and becomes positive since the corrections for the dimensional effects given in (6.28) are positive and increase as $\sim R^{-3}$.

Thus, when passing to small whisker diameters, the corrections for the dimensional effects reverse the signs of all the coefficients in (6.1). In this case the new equilibrium state will be the state of the unstable focus-type.

Nonformally the mechanism of positive feedback occurrence in the system can be explained by the dimensional dependence of the growth angle. For the section of thin whisker contraction the growth angle ψ_k increases a little faster than the slope of the line tangent to the melt meniscus. In this case, the smaller the whisker diameter at the drop base, the faster it should contract. However, simultaneously oversaturation decrease occurs at the crystallization front, which results in decrease of the crystallization rate and the rate of whisker contraction. The dynamic equilibrium is unstable and develops in the opposite direction as soon as the process is stopped.

6.4. Profile Curves of Initial Whisker Sections

The process of whisker growth is as follows. When a metal particle is melted into the substrate a melt drop assumes the shape of a lens. The lens geometry is determined by the condition of surface-tension force equilibrium at some point on the three-phase line (Figures 1.7, 6.3b). Since the gas mixture flows above the melt drop and the oversaturation required is achieved in the melt, crystallization initiation is observed at the front, the drop base representing this front. If in an isotropic approximation the meniscus makes an angle ψ with the horizontal surface of the substrate (Figure 6.3b 1), in the process of crystallization the angle between the line tangent to the meniscus and the horizontal increases, the crystallization front coincides with the substrate surface at some instant. In this case $\psi = \psi_0$ (Figure 6.3b 2). From this instant the initial cone-like whisker section crystallizes (Figure 6.3b 3).

Ignoring the dimensional effects for rather thick whiskers, the growth angle can be considered as a constant at any instant of crystallization. Then the process of whisker growth follows the scheme of small-drop crystallization described in 1.3.3. The only difference is that during drop crystallization the overall crystal and melt mass is constant while during whisker crystallization the volume of the meniscus on the whisker end-face can be regarded as constant.

The profile curves of the initial sections were numerically calculated and are given in (Figure 6.3a). The initial data are the experimentally-found angle $\psi_0 \simeq 54^0$

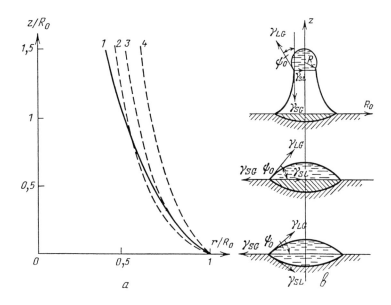

Fig. 6.3.   On calculating the initial whisker section: 1 – a whisker profile curve taken from a photograph in [135]; 2, 3, 4 – calculated curves for $\psi_0 = 54°$, 64° and 7-°, respectively (a); diagram illustrating formation of the initial cone-shaped section (b).

[138] and the drop volume W selected arbitrarily. The initial dimension R_0 of the whisker base is calculated from the formula $R_0 = [6W\pi D(3 + D^2)]^{1/3}$, where $D = (1 - \cos\psi_0)\sin\psi_0$. Some discrepancy of the experimental and calculated curves can be attributed to the inaccuracy of growth angle measurements as well as to the simplified scheme of calculation ignoring changes in the drop volume in the process of crystallization front straightening as the initial whisker section grows.

Note that the initial-cone formation results in the first necking both for thick and thin silicon whiskers. This fact is attributed to the first oscillation in the process of transition to the stationary state of the stable focus-type for thick whiskers or to the conditions of self-excited oscillations for thin whiskers.

In conclusion it should be noted that the mechanism offered here does not reject other possible models interpreting radial instability in the process of whisker growth. Lack of discrepancy of the model offered is demonstrated here. It is quite obvious that to finally solve this problem a study quantitatively verifying the conditions under which self-excited oscillations occur and not for silicon alone is required.

Chapter 7

CYLINDRICAL PORE GROWTH

7.1. Formulation of the Problem

Figure 7.1. shows stationary growth of cylindrical pores. Here the meniscus can be considered as a sphere since the pore diameter is equal to a few hundred micrometers-small Bond number. The pore grows stationary if a line tangent to the meniscus makes the angle of growth with the normal to crystallization front.

If the pore diameter $2R$ and the gas concentration C within a pore are regarded as variable parameters the set of equations (1.3) will include (1.39) and the equation of the concentration change rate:

$$\delta \dot{R} = A_{RR}\delta R + A_{RC}\delta C \tag{7.1}$$

$$\delta \dot{C} = A_{CR}\delta R + A_{CC}\delta C. \tag{7.2}$$

The pressure P within a pore can be considered equal to CBT, where B is the universal gas constant, T is the absolute temperature.

7.2. The Capillary Coefficients A_{RR} and A_{RC}

From equation (1.39):

$$A_{RR} = -V\frac{\partial \alpha_0}{\partial R}$$

and

$$A_{RC} = -V\frac{\partial \alpha_0}{\partial C},$$

where $\alpha_0 = \pi/2 - \psi_0$ and V is the rate of crystal growth.

From Figure 7.1. the following ratio can be derived:

$$\frac{R}{r} = \sin \alpha_0. \tag{7.3}$$

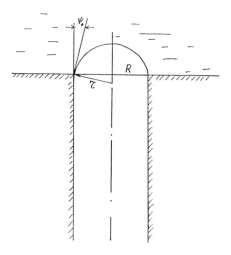

Fig. 7.1.   Stationary growth of cylindrical pore.

Here r is the radius of curvature of the spherical meniscus.

The Laplace capillary equation has the following form in this case:

$$\frac{2\gamma}{r} = CBT. \tag{7.4}$$

Simultaneous solution of Equations (1.39), (7.3) and (7.4) gives the explicit functions A_{RR} and A_{RC}:

$$A_{RR} = -\frac{V}{r\cos\alpha_0} = -\frac{V}{R}\mathrm{tg}\alpha_0 = -\frac{VCBT}{(4\gamma^2 - R^2C^2B^2T^2)^{1/2}} \tag{7.5}$$

$$A_{RC} = -\frac{VRBT}{2\gamma\cos\alpha_0} = -\frac{VCBT}{(4\gamma - R^2C^2B^2T^2)^{1/2}}. \tag{7.6}$$

7.3.   The Concentration Coefficients A_{CR} and A_{CC}

The rate of gas-concentration changing in a pore depends on the rate of the gas in flow through the surface of the melt meniscus and the rate of pore-volume changing:

$$\frac{dC}{dt} = \frac{1}{W}\left(\frac{dm}{dt} - C\frac{dW}{dt}\right). \tag{7.7}$$

Here m is the mass of the gas in the pore, W is the pore volume: $W = \pi R^2 l$, where l is the pore length. The explicit function of the rate of the gas inflow through the surface of the melt meniscus can be written down as follows:

$$\frac{dm}{dt} = \beta(C_m - C)s. \tag{7.8}$$

Here β is the kinetic coefficient that will be discussed below, C_m - the concentration of gas in the melt on the surface of the liquid meniscus, S is the area of the liquid meniscus (a part of the spherical segment):

$$S = 2\pi r(r - (r^2 - R^2)^{1/2}). \tag{7.9}$$

The rate of pore volume changing can be given by the following equation:

$$\frac{dW}{dt} = \pi R^2 V. \tag{7.10}$$

In accordance with (1.3) the coefficients A_{CR} and A_{CC} are defined by differentiating the right side of Equation (7.7):

$$A_{CR} = \frac{\pi R}{W} \left[\frac{4\beta\gamma(C_m - C)}{(4\gamma^2 - C^2 B^2 T^2 R^2)^{1/2}} - 2CV \right] \tag{7.11}$$

$$A_{CC} = \frac{\pi}{W} \left((C_m - C) \left[\frac{4\beta R^2 \gamma}{C(4\gamma^2 - C^2 B^2 T^2 R^2)^{1/2}} - \frac{16\beta\gamma^2}{\beta^2 C^3 T^2} 2CV \right. \right.$$
$$\left. \times \left(1 - \left(1 - \frac{R^2 C^2 B^2 T^2}{4\gamma^2} \right)^{1/2} \right) \right]$$
$$\left. - \frac{8\beta\sigma^2}{B^2 T^2 C^2} \left(1 - \left(1 - \frac{R^2 C^2 B^2 T^2}{4\gamma^2} \right)^{1/2} \right) - R^2 V \right). \tag{7.12}$$

7.4. Analysis of the Process-Stability Conditions

The process-stability conditions will be analyzed for $\psi_0 = 14^0$ that is characteristic for some insulators, e.g. sapphire.

$$A_{RR} = -\frac{4V}{R} \tag{7.13}$$

$$A_{RC} = -VR\frac{2BT}{\gamma} = -\frac{4V}{C} \tag{7.14}$$

$$A_{CC} = \frac{1}{l} \left[\beta(5\frac{C_m}{C} - 6.6) - V \right] \tag{7.15}$$

$$A_{CR} = \frac{2C}{Rl} \left[4\beta \left(\frac{C_m}{C} - 1 \right) - V \right]. \tag{7.16}$$

Now from the Routh-Gurvitz conditions ($A_{RR} + A_{CC} < 0$; $A_{RR}A_{CC} - A_{RC}A_{CR} > 0$) the necessary and sufficient conditions of the growth of cylindrical ($l \gg R$) pores can be defined:

$$3\frac{C_m}{C} - 1.5 > \frac{V}{\beta} > \frac{5}{4}\frac{R}{l}\left(\frac{C_m}{C} - 1.3\right). \tag{7.17}$$

In particular, for the beginning of the process ($l \sim 2R$) the conditions are as follows:

$$3\frac{C_m}{C} - 1.5 > \frac{V}{\beta} > \frac{5}{9}\frac{C_m}{C} - 0.73. \tag{7.18}$$

From [139] it follows that $\beta \sim \frac{D_l}{R}$, where D_l is the diffusion coefficient of the gas in the melt, that is equal to $10^{-5}\text{cm}^2\text{s}^{-1}$. It allows us to rewrite conditions (7.18) in the following form:

$$10^{-5}\left(\frac{3}{2}\frac{C_m RBT}{\gamma} - 1.5\right) > RV > 10^{-5}\left(\frac{1}{4}\frac{C_m RBT}{\gamma} - 0.75\right). \tag{7.19}$$

Here we did not use the Gibbs-Thomson dimensional dependence of growth angle as in the previous chapter. But as a first approximation this model helps to increase our understanding of the conditions of sufficiently big diameter (more than some micrometers) cylindrical pore appearance.

Chapter 8

TWO SHAPING ELEMENTS TECHNIQUE

8.1. Capillary Shaping

To define specific features of capillary shaping of plates in crystallization using two shaping elements (TSET) it is necessary to solve the capillary problem for Equation (1.13) with the catching boundary conditions at the feeder free edge and the angle fixation condition at the crystal free edge (Figure 1.10). Profile curves for various pressure values for crystals 0.25 mm thick are given in Figure 8.1.

The fact that in the growth of thin-walled crystals the meniscus length is an order of magnitude higher than when one shaping element, a shaper, is used is immediately noticeable. The second specific feature is that the meniscus length does not decrease with the growing crystal thickness decreasing as when a shaper is used, but in contrast increases.

The shaping function of the substrate is primarily to 'support' the meniscus end-face sections where its length is smaller than in its main part. Furthermore, the substrate presence allows large deviations of the crystallization front from a straight line, which extends the technological potential of the technique.

8.2. Hydrodynamic Conditions

At first, the process of flat horizontally pulled substrate wetting by the melt will be considered without an allowance for crystallization. The solution of this problem will provide data on minimum film thicknesses. A similar problem regarding liquid entrained by an infinite plate vertically pulled out of it was analyzed in [140]. Let the liquid surface on the substrate be divided into two sections: static meniscus I limited by the feeder free edge and substrate-entrained liquid film II (Figure 8.2.).

Range I is described by the Laplace capillary equation (1.13), range II is described by the Navier-Stokes equation as a boundary-layer approximation. From the condition of solution joining for ranges I and II (continuity of d^2h/dr^2) the amount of substrate-entrained liquid and the liquid-film thickness h_f can be determined.

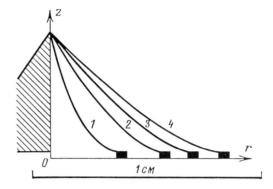

Fig. 8.1. Profile curves for various pressures: $1 - 1$; $2 - 0.5$; $3 - (-0.25)$; $(4) - (-0.1)$.

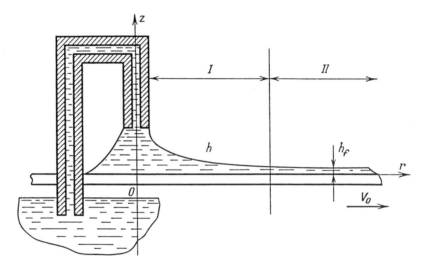

Fig. 8.2.   Hydrodynamic conditions of crystallization using a feeder and substrate: 1 – the range of capillary shaping; 2 – the range of viscous flow; h_f – film thickness; V_0 – the rate of substrate motion.

So, range II is described by the following equation:

$$\eta \frac{d^2 V}{dz^2} = \frac{1}{\rho} \frac{dP}{dr} \tag{8.1}$$

with the following boundary conditions:

$$V = V_0, \quad \text{if} \quad z = 0 \tag{8.2}$$

$$\frac{dV}{dz} = 0, \quad \text{if} \quad z = h. \tag{8.3}$$

Here η is the kinematic melt viscosity, V is a component of pulling rate parallel to the pulling direction, P denotes the capillary pressure, V_0 is the substrate rate, z is the vertical coordinate measured from the substrate surface, $h(r)$ is the melt surface profile.

Integration of (8.1) gives

$$V(z) = V_0 - \frac{1}{\eta}\frac{dP}{dr}\left(hz - \frac{z^2}{2}\right). \tag{8.4}$$

The melt flow in the film can be defined from the continuity equation

$$j_0 = V_0 h_f = V_0 h - \frac{1}{3\eta}\frac{dP}{dr}h^3 \tag{8.5}$$

where j_0 denotes the melt flow per unit substrate width and $h_f = \lim h$ is the liquid-film thickness at a large distance from the feeder. Taking into account that the film curvature is small, it can be considered that

$$\frac{dP}{dr} = \frac{d}{dr}\left(-\frac{d^2h}{dr^2}\right) = -\gamma_{LG}\frac{d^3h}{dr^3}. \tag{8.6}$$

Substitution of (8.6) into (8.5) and solution of the equation obtained give

$$\frac{d^3h}{dr^3} = \frac{3\eta}{\gamma_{LG}}\left(\frac{j_0}{h^3} - \frac{V_0}{h^2}\right). \tag{8.7}$$

Now the following dimensionless coordinates will be introduced:

$$H = \frac{h}{h_f}, \qquad X = \left(\frac{3\eta V_0}{\gamma_{LG}}\right)^{1/3}\frac{r}{h_r}. \tag{8.8}$$

Then Equation (8.7) is rearranged into the following form

$$\frac{d^3H}{dX^3} = \frac{1}{H^3} - \frac{1}{H^2} \tag{8.9}$$

with the boundary conditions: $H \to 1$ for $X \to \infty$, $H \to \infty$ for $X \to 0$,

$$\frac{dH}{dX}\frac{d^2H}{dX^2} \to 0 \qquad \text{for} \qquad X \to \infty.$$

Equation (8.9) is similar to that derived in [140], the difference being that the contribution of the gravity force has not been overlooked, hence the equation is more accurate. In [140] it is shown that

$$\frac{d^2H}{dX^2}\bigg|_{H\to\infty} \to \nu = 0.63. \tag{8.10}$$

The range of static meniscus is described by Equation (1.13):

$$\frac{d^2h/dr^2}{[1 + (dh/dr)^2]^{3/2}} = \frac{\rho g}{\gamma_{LG}}(P + z) \tag{8.11}$$

crystallization technique	A_{RR}, s^{-1}	A_{Rh}, s^{-1}	A_{hR}, s^{-1}	A_{hh}, s^{-1}	τ_R, s^{-1}
TPS	-0.044	-0.070	0.970	-0.085	15.5
TSET	-0.230	-0.006	90.45	-0.112	5.8

$\psi_0 = 11^{\circ}$, $V = 3.3 \ 10^{-4} \ \text{m s}^{-1}$, $P = 10^{-2} \text{m}$, $\mathcal{L} = 4.61 \ 10^{9} \ \text{J} \cdot \text{m}^{-3}$,

$\rho = 2560 \ \text{kg m}^{-3}$, $\lambda = 60 \ \text{wt m}^{-1}$, $\lambda_s = 21.6 \ \text{wt m}^{-1}$, $T_0 = 1685 \ \text{K}$,

$T_1 = 1385 \ \text{K}$, $\mu_{rad} = 414 \ \text{wt m}^{-2}$, $\mu_{conv} = 10 \ \text{wt m}^{-2} \text{K}^{-1}$

TABLE 8.1

with the boundary condition $\frac{dh}{dr} = 0$ for $z = 0$, hence

$$\frac{d^2 h}{dr^2} = \frac{\rho g}{\gamma_{LG}} P.$$

The passing to the dimensionless coordinates (8.8) gives:

$$\frac{d^2 H}{dX^2} = \frac{\rho g h_f}{(3\eta V_0)^{2/3} \gamma_{LG}^{1/3}}. \tag{8.12}$$

From (8.10) and (8.12) it follows that

$$h_f = 1.31 \frac{(\eta V_0)^{2/3} \gamma_{LG}^{1/3}}{\rho g P}. \tag{8.13}$$

Substitution of the values of the silicon melt constants (Table 8.1) into (8.13) gives:

$$h_f = 0.39 V_0^{2/3} P^{-1}. \tag{8.14}$$

Taking into consideration the substrate displacement rates within the range used in the growth practice (1–5 cm/min) and pressures providing spontaneous filling of capillary channels, Equation (8.14) gives the values of liquid film thickness h_f within the interval of 0.1 to 3 μm. Thus, the capillary pressure gradient along the r-axis inhibits melt entrainment by the substrate caused by viscosity forces, the amount of substrate-entrained melt being small due to considerable surface tension and low viscosity of melted silicon.

The calculations made lead to the consideration of two possible regimes of melt crystallization on the substrate. Where the heat conditions specify the crystallization front position within liquid-film range II (Figure 8.2), the thickness of the layer obtained does not depend on R, the front coordinate, and hence on its shape. Such crystallization conditions are attractive from the point of view of growth procedure because the heat conditions need not be controlled to produce a crystalline layer uniform along the substrate width. However, such conditions can be used only in cases where very thin layers are to be grown (another case when crystallization in range II can be used is connected with utilization of a melt-wettable porous substrate; it will be discussed below).

To produce various devices, photoconverters in particular, a silicon layer $\sim 100\mu$m thick is required; in this case the crystallization front should lie within the range of static meniscus as is shown in Figure I.11. Under such conditions the thickness of the growing crystal depends on the meniscus height at the point of its intersection with the crystallization front and the corresponding boundary conditions of angle fixation on the layer side and meniscus catching on the operating feeder free edge are satisfied. Thus, crystallization in range I requires controllable heat conditions, i.e., it states the problem of matching the heat and the capillary conditions of crystallization usual for melt growth of crystals.

8.3. Stability Investigation

A model of TSET regarded as a system of two degrees-of-freedom, R and h, can be constructed (Figure I.10). While determining the coefficient A_{hR}, heat removal is considered to take place from the upper surface of the crystallizing layer while the carbon fiber-reinforced substrate impregnated with the melt is a heat insulator.

Table 8.1. gives the values of the coefficients of the stability equations. Figure 8.3 gives calculated curves of attenuations of crystallization front position perturbations $\delta h(f)/\delta h(0)$ for silicon tapes grown by TPS and by TSET under similar conditions. The behavior of the attenuation curves and the values of the main stability parameter τ_{rp} (the time of perturbation relaxation) show higher stability of the TSE-technique. This result agrees with experimental data.

8.4. Impurity Distribution

The process of impurity transfer from the bulk of the melt towards the growth meniscus and impurity transfer within the meniscus will be considered in accordance with an idealized diagram given in Figure 8.4. Melt contamination by impurities of the substrate material is not taken into consideration here.

Let us assume that in the melt zone between the feeder and the substrate complete stirring takes place, while in the meniscus a laminar liquid flow is observed for which impurity transfer is described by the one-dimensional diffusion equation

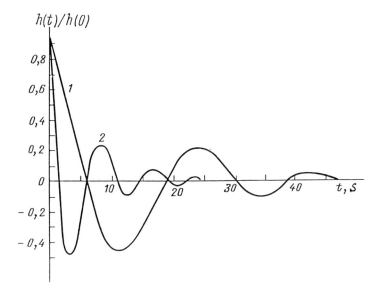

Fig. 8.3.   Calculated curves of attenuations of crystallization front perturbation for a silicon tape; 1 –
TPS; 2 – TSET.

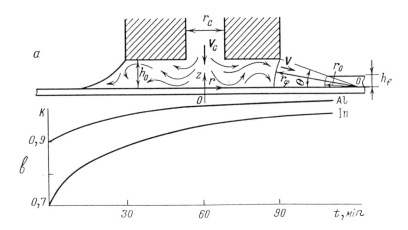

Fig. 8.4.   On calculating impurity distribution in the process of crystallization using a feeder and a
substrate (a) and calculated curves of time dependence of the effective coefficients K of aluminum and
indium impurity distribution in silicon (b); r_c is the size of the feeding channel.

$$D \left(\frac{d^2C}{dr^2} + \frac{1}{r} \frac{dC}{dr} \right) = -V \frac{dC}{dr} \tag{8.15}$$

with the following boundary conditions at the crystallization front:

$$-D\frac{dC}{dr}\Big|_{r=R} = -V_0(1 - K_0)C(R) \tag{8.16}$$

and at the left meniscus boundary

$$C\big|_{r=R} = C_m. \tag{8.17}$$

Here D is the coefficient of impurity diffusion in the melt, C_m denotes impurity concentration in the stirring zone, other notations are given in Figure I.11.

The rate field of the melt flow in the meniscus is specified by the relation $V = V_0 r_0/r$; allowing that Equation (8.15) takes the following form:

$$\frac{d^2C}{dr^2} + \frac{1}{r}\left(1 + \frac{V_0 r_0}{D}\right)\frac{dC}{dr} = 0. \tag{8.18}$$

Solution of this equation with the boundary conditions (8.16) and (8.17) allows us to derive a ratio K^* of the impurity concentration in the crystal C_S to its concentration in the stirring zone C_m ($K^* = C_S/C_m$):

$$K^* = \frac{K_0}{K_0 + (1 - K_0)(r_0/r_\varphi)^{\alpha*}} \tag{8.19}$$

where $\alpha_* = V_0 r_0/D$, K_0 is the equilibrium coefficient of impurity distribution.

Substitution of simple geometrical ratios $r_0/r_\varphi = h_f/h_0$, $r_0 = h_f/\sin\theta$, reduces Equation (8.19) to the following form:

$$K^* = \frac{K_0}{K_0 + (1 - K_0)(h_f/h_0)^{h_f V_0/D \sin\theta}}. \tag{8.20}$$

Since for the crystallization scheme under consideration the length of the feeder capillary channel l_c is large in comparison with the length of the shaper capillary channel of TPS, the condition $V_c \gg D/l_c$ (V_c is the rate of the laminary melt flow through a capillary channel of the width r_c (Figure 8.4) holds for the most part of the impurities in silicon melt and the reverse impurity diffusion into the bulk of the melt can be ignored.

Impurity concentration in the stirring zone C_m depends on time in the following manner:

$$C_m(t) = C_\infty + \frac{V_c r_c C_\infty - K_0 C(r_0)}{W_m}t \tag{8.21}$$

where W_m denotes the stirring-zone volume, t is the time. With an allowance for the effective coefficient of impurity distribution $K = C_S/C_\infty$ Equation (8.21) allows to derive an expression for $K(t)$:

$$K(t) = \frac{1 + V_c r_c t/W_m}{1/K^* + V_0 h_f t/W_m}. \tag{8.22}$$

For the most part, impurities in silicon melt $K \approx 1$ irrespective of the growth time, however for impurities characterized by small K_0-values and large D-values the effective coefficient of impurity distribution reaches the value of *one* only for a long period of time.

Figure 8.4. shows the values of $K(t)$ for Al ($K_0 = 0.002$, $D = 5.3 \ 10^{-8} m^2 s^{-1}$) and In ($K_0 = 0.0004$, $D = 5.2 \ 10^{-8} m^2 s^{-1}$). The values of $h_f = 2.5 \ 10^{-4} m$, $r_c = 8 \ 10^{-4} m$, $W_m = 1.16$ cm^3, $h_0 = 3.5 \ 10^{-3} m$, $V_0 = 3 \ 10^{-4} m \ s^{-1}$ used in practice are taken.

The analysis carried out demonstrates that macrouniform crystalline layers can be grown by silicon crystallization using a feeder and a substrate. The effect of gradual approach of K to *unit* becomes important only for thin-layer growth when crystallization takes place in liquid-film range II (Figure 8.2.)

8.5. Application of Carbon Fiber-Reinforced Substrates

For the first time in the practice of silicon profiling tapes of fibrous carbon materials (FCM) in the form of woven fabric were used as a substrate in this technique.

Application of FCM as a substrate material allows the use of carbon fabric that can be fed to the melt-treatment zone from a compact coil, that is easily cut into preforms of desired shapes, is commercially available in the form of woven fabric of various thickness and weaving and provides deepcleaning of FCM from metal impurities intolerable in silicon of semiconductor application (since FCM developed surface promotes their effective cleaning by halogen-containing agents).

Carbon-fabric tape is siliconized when in contact with the melt, it is impregnated with the melt, some part of the fiber materials being transformed into silicon carbide. The heterogeneous ternary material obtained assumes totally new properties, therefore it is more correct to consider carbon fabric not only as a substrate but also as a precursor for manufacturing carbon-silicon composites using carbon fibers. This material is conventionally called silicon-on-fabric (SOF) [41], the abbreviation SOF being used from now on to denote both the component itself and the process of silicon crystallization using two shaping elements, a dense-graphite feeder and a FCM-based substrate.

Carbon-fabric impregnability with silicon melt allows a few versions of the SOF-process (Figure 8.5.)

$I - N$ version (impregnation of N densely packed tapes). Crystallization takes place at a sufficient distance from the feeder (in range II, Figure 8.2.)

The S-version (sandwich) provides manufacture of plates 2–6 mm thick by crystallization of the melt inside the clearance between two tapes, the melt being pulled through the crystallization zone. In this case the amount of carbon in a rather thick SOF plate can be reduced to a minimum (down to 0.05 mass %), arbitrary positioning and crystallization front shape being kept, which is an important advantage of crystallization in range II.

The L-version (layer) is used when the crystallization of a semiconducting crys-

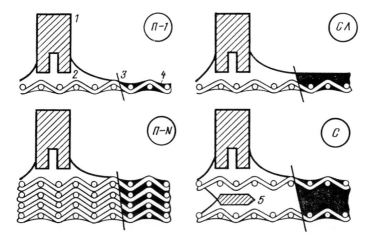

Fig. 8.5. Versions of the SOF-process different in the heat and capillary conditions of crystallization, the number of carbon-fabric tapes and the technique of their positioning: 1 – feeder; 2 – meniscus; 3 – crystallization front; 4 – silicon layer; 5 – separator, $I-1$ impregnation of one tape, $I-N$ impregnation of N tapes, L Si – layer crystallization; P Si – plate manufacture.

talline layer grows directly from the melt on an electroconductive substrate in the course of one production cycle.

8.6. Silicon as a Component of the SOF Material

Crystalline structure of free silicon as a component of the SOF material is of greatest importance for the SOF $-L$ material since the grain size of polysilicon affects photoconverter efficiency.

Figure 8.6a shows a SOF $-L$ specimen. The silicon layer is formed by elongated grains 0.5-3 mm wide and 5-30 mm long stretched in the direction of growth. Here and there multi-twinned sections whose structure is similar to that of thin-walled profiled silicon crystals grown by TPS (see 11.5) can be observed but their share does not exceed 3-5% of the overall area.

In the immediate vicinity of the carbon-fabric filaments fine, practically equiaxial grains 20-50 μm in size can be observed (Figure 8.6b). The whole of the silicon phase in the SOF $-I-N$ materials is formed by similar crystals. The patterns of the SOF $-L$ transverse microsection metallographic specimen given show that the boundaries of coarse grains forming the surface silicon layer are preferentially oriented perpendicular to the surface.

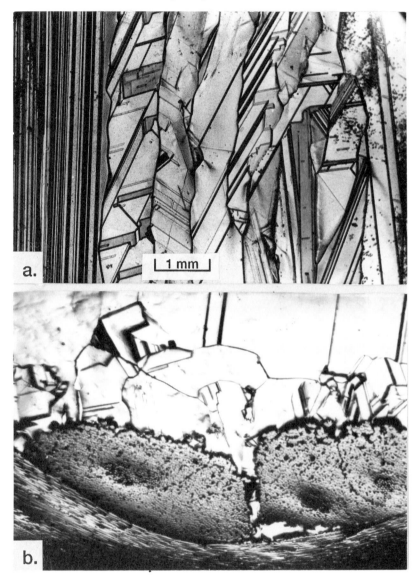

Fig. 8.6.   Etched surface (a) and cross-section (b) of SOF-specimens.

8.7. Prospects of Carbon-Silicon Material Application

High output of the SOF-process combined with low costs of precursor cleaning creates prerequisites for commercial application of SOF-materials of various types.

SOF $-I-N$, SOF $-S$ materials can be used as structural ones to produce elements of high-temperature equipment for microelectronic application (substrate-holding cassettes, screens, tubes, etc.). Quartz glass currently used for these purposes is expensive, not high-temperature resistant, and represents a source of undesirable impurities.

SOF $-I-N$ materials are good for heaters used in air at temperatures of up to 1200°C.

SOF $-L$ materials are designed for photoconverter substrate application. Usually, silicon layer resistivity does not exceed 1 Ohm cm, resistivity of the ternary Si-SiC $-C$ base is 0.02-0.04 Ohm cm.

Thus, the SOF-technique enables the production of a semiconducting crystalline layer on an electroconductive substrate directly from the melt in the course of one production cycle.

Chapter 9

VARIABLE AND LOCAL SHAPING TECHNIQUES

The investigation of the dynamic stability of the variable shaping technique does not differ from that of TPS. It is only necessary to analyze the stability of each part of the crystal unchanged cross-section. The scheme of analysis described in Chapter 3 can be used.

As regards the local shaping technique, in fact it is the combination of TPS and TSET, where the bottom of the grown crystal plays the role of substrate.

Let us analyze the dynamic stability of the local shaping technique using the example of tube growth (Figure I.13) [50].

9.1. Formulation of Problem

In steady-state growth, the length of the crystal increases by $l = V/n$ per revolution. From the viewpoint of mass balance, this is equivalent to a situation as though, at the interface, there were a macrostep normal to the growing surface (Figure 9.1). The height to the step is $l = V/n$, its growth occurs in the plane of the step only, and there is no mass transfer over the rest of the melt-crystal interface. Therefore, as in the analysis of other processes of growing crystals from the melt, the real shape of the crystallization front can be replaced by a planar front: in our case, the flat surface of the step is assumed to serve as the crystallization front. The position of the crystallization front relative to the edge of the die will be given by the displacement $u = R\phi$, where ϕ is the central angle, measured with respect to the shaper's edge in the direction of the crystal rotation.

If the level of melt in the crucible is assumed to be constant during the course of crystallization, the process will have three degrees-of-freedom (i.e., three independent variables): the height of the step l; the angle ϕ, which determines the position of the crystallization front relative to the shaper's edge; and the wall thickness, m, of the tube. However, since the pulling rate of the tube is much less than its linear rate of rotation $(V/2\pi Rn~10^{-3}$ to $10^{-4})$, then for perturbations of steady-state growth, the thickness of the wall will vary much more slowly than the height of the step l. This

223

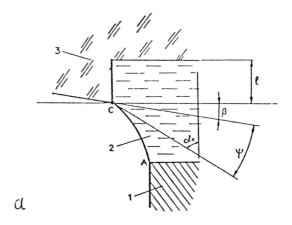

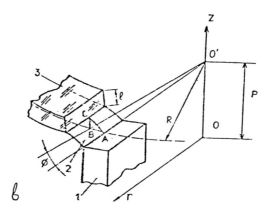

Fig. 9.1.   Schematic representation of crystal formation (a) and degrees-of-freedom (b): 1 – shaper; 2 – meniscus; 3 – crystal.

allows an analysis with two degrees-of-freedom only: l and ϕ. The linearized set of equations (1.3) is of the form:

$$\delta \dot{l} = A_{ll}\delta l + A_{l\phi}\delta \phi \qquad\qquad (9.1)$$

$$\delta \dot{\phi} = A_{\phi l}\delta l + A_{\phi\phi}\delta \phi \qquad\qquad (9.2)$$

9.2. Capillary Coefficients A_{ll} $A_{l\phi}$

The estimates made in 1.2. have shown that, at rates of crystallization below 10 to 100 cm/s, the effects of inertial forces can be neglected. The rate of crystallization of sapphire tubes of 60 mm diameter does not exceed 2 cm/s, which makes it possible to consider the formation of the melt column in a hydrostatic approximation.

When stationary growth is disturbed, the rate of variation of the step height is similar to that of the characteristic transverse size of crystals in other techniques of crystallization from the melt, and may be given by the relation (1.38) $dl/dt = V\delta\alpha_0 = 2\pi Rn\delta\alpha_0$, where $\delta\alpha_0 = \alpha_0 - \alpha_e$, is the angle between the tangent to the meniscus at point C and the vertical (Figure 5), and α_e is the equilibrium value of the angle α_0. In steady-state growth $\alpha_0 = 1/2\pi - \psi_0 - \beta$, where ψ_0 is the growth angle and β is the angle of the spiral line slope. Since tg$\beta \approx \beta \approx V/2\pi Rn = 10^{-3}$ to 10^{-4}, one may neglect its value and assume that $\alpha_e = 1/2\pi - \psi_0$. The linearized equation for the rate of step height variation is

$$\delta\dot{l} = -2\pi Rn\frac{\partial\alpha_0}{\partial l}\delta l - 2\pi Rn\frac{\partial\alpha_0}{\partial\phi}\delta\phi \qquad (9.3)$$

where $A_{ll} = -2\pi Rn(\partial\alpha_0/\partial l)$ and $A_{l\phi} = 2\pi Rn(\partial\alpha_0/\partial\phi)$. The values of the derivatives $\partial\alpha_0/\partial l$ and $\partial\alpha_0/\partial\phi$ are determined from the solution of the Laplace capillary equation. In the present case, the meniscus is not symmetrical and is rather complex in shape. However, to analyze the stability of the crystallization process qualitatively (as was done for TSET), the sign of the derivatives can be determined from the shape of the profile curve AC, without considering the bulk of the melt column. In solving the Laplace equation, the boundary condition of catching the meniscus at the edge of the shaper and the condition of constancy of the growth angle are used.

We consider two limiting cases: (1) AC is the profile curve of the meniscus formed in the process of growing a plate; and (2) AC is the profile curve of the meniscus formed in the process of growing a thin rod with radius $r_c < a$, where a is the capillary constant. Using the results of detailed studies of shapes of different menisci and of the corresponding capillary coefficients (see Chapter 3), it can be easily shown that, under the given boundary conditions, we have $A_{ll} < 0$ and $A_{l\phi} < 0$ for both limiting cases considered.

9.3. Thermal Coefficients $A_{\phi l}$ and $A_{\phi\phi}$

The values of the thermal coefficients can be found from the condition of heat balance at the crystallization front (1.44).

In this case, $V_c = V - du/dt = 2\pi Rn - R(d\phi/dt)$. The temperature gradients (G) at the crystallization front, as in the analysis of other methods of crystallization from the melt are determined from the solution of the one-dimensional heat conductivity equation by assuming the tube wall thickness $m \gg l$. The boundary conditions involve the given temperatures at the crystallization front, T_0, and that of the melt, T_m, at a distance of $u = R\phi_0$ from the crystallization front, where ϕ_0 is the value of angle determining the position of the step l_0 under steady-state conditions. Far from the crystallization front, the temperature gradient in the solid phase is taken to be *zero*. It follows from the solution that $G_S = -T_0\xi$ and $G_L = (T_0 - T_m)/R\phi_0$, where

$$\xi = \frac{VR}{2k_s} - \left[\left(\frac{VR}{2k_s}\right)^2 + \frac{\mu_s R^2}{\lambda_s l}\right]^{1/2}.$$

k_s and μ_s are the coefficients of thermal conductivity and heat exchange, respectively, of the solid phase. Taking into account the fact that

$$VR/2k_s \ll \mu_s R^2/\lambda_s l$$

we obtain

$$\xi = (\mu_s R^2/\lambda_s l)^{1/2}.$$

Substituting the values of temperature gradients and crystallization velocity into Equation (1.44) and linearizing the latter, we obtain

$$A_{\phi l} = \frac{\lambda_s T_0}{2\mathcal{L}\rho R}\frac{\xi}{l} > 0, \quad A_{\phi\phi} = \frac{\lambda_L T_0 - T_m}{\mathcal{L}\rho R^2 \phi_0^2}. \tag{9.4}$$

It follows that, when the melt is overheated, $T_0 < T_m$ and $A_{\phi\phi} < 0$. When the melt is supercooled, $A_{\phi\phi} > 0$.

9.4. Analysis of Stability

In order for the set of equations (9.1) and (9.2) to be stable, it is necessary and sufficient that (Routh-Gurvitz conditions):

$$A_{ll} + A_{\phi\phi} < 0 \tag{9.5}$$

$$A_{ll}A_{\phi\phi} - A_{l\phi}A_{\phi l} > 0. \tag{9.6}$$

From the values of the coefficients obtained in the foregoing analyses, we can conclude that the conditions (9.5) and (9.6) are satisfied when the melt is overheated. The experimental data indicate that the surface of the crystallized layer melts whenever it enters the melt column, thus indicating overheating of the melt. Therefore, we may conclude that the growth of shaped crystals by the local shaping technique is realized in practice as a result of the stability of the process.

Chapter 10

CRYSTAL CROSS-SECTION SHAPES

In the previous chapters, uniform perturbations (change in the radius of a cylindrical crystal, tube or in plate thickness) were superimposed on the cross-section of a growing crystal when stability was analyzed. Such a consideration solves the problem of characteristic profile dimension stability, however the problem of maintaining the shape of a crystal being pulled remains unsolved.

To analyze shape stability, an arbitrary small perturbation that can be expanded into a series with respect to a corresponding set of orthogonal functions depending on the shape of the perturbed contour, should be superimposed on the contour limiting the profile cross-section and then the time dependence of the coefficients of this expansion should be calculated. In this case, the cross-section shape is stable provided the coefficients of all the terms of the expansion decrease with time. The first term of the expansion corresponds to the uniform perturbation, i.e., to some change in the cross-section dimension while its shape remains unchanged. The unperturbed meniscus shape described by the function z_0 satisfying the Laplace capillary equation with corresponding boundary conditions and the equilibrium angle α_e corresponds to an unperturbed contour. The solution $z = z_0 + z_1$ satisfying the two-dimensional equation (1.9) linearized in respect to a small correction factor z_1 and the corresponding angle $\alpha_0 = \alpha_e + \delta\alpha_0$ will correspond to the small perturbation of the contour shape. Since all the equations obtained in this case are linear with respect to z_1, the behavior of an arbitrary small and rather smooth distortion can be investigated by analyzing the behavior of the superposition of the components of z_1 expansion with respect to the corresponding set of orthogonal functions.

10.1. Plate-Shaped Crystal Pulling by TPS [141]

The Laplace capillary equation in terms of dimensionless parameters for an unperturbed contour in the Cartesian coordinate system given in Figure 10.1. takes the form of (1.13):

$$z_0'' + 2z_0(1 + z_0'^2)^{3/2} = 0 \qquad (10.1)$$

227

where z_0 is the function of x alone. Let the catching boundary condition be satisfied at the shaper free edge:

$$z_0 \mid_{x=x_0} = d. \tag{10.2}$$

The crystallization front position is determined by the catching condition at the crystal free edge:

$$z_0 \mid_{x=a_0} = h. \tag{10.3}$$

Here $2x_0$ is the shaper dimension, $2a_0$ denotes the plate thickness. The angle between the horizontal and the line tangent to the meniscus at the crystallization front is equal to its equilibrium value α_e:

$$z_0' \mid_{x=a_0} = P \mid_{x=a_0} = P_0 = -\text{tg}\alpha_e.$$

Here the notation is introduced $dz_0/dx = P$. The solution of the perturbed problem can be represented in the following form:

$$z = z_0(x) + z_1(x, y)$$

where $z_0(x)$ satisfies (10.1)–(10.3). Substitution of this solution into the two-dimensional capillary equation (1.9) with only first-order terms with respect to z_1 left gives it a linearized equation for $z_1(x, y)$:

$$\frac{\partial^2 z_1}{\partial x^2} - \frac{\partial z_1}{\partial x} \frac{6 z_0 P}{1 + P^2} (1 + P^2)^{3/2}$$

$$-2 z_1 (1 + P^2)^{3/2} + \frac{\partial^2 z_1}{\partial y^2} (1 + P^2) = 0 \tag{10.4}$$

with the following boundary conditions:

$$z = (z_0 + z_1) \mid_{x=x} = d \tag{10.5}$$

$$z = (z_0 + z_1) \mid_{x=a(y)} = h \tag{10.6}$$

where $x = a(y)$ is the perturbed-contour equation. The solution of (10.4) can be represented in the following form:

$$z_1(x, y) = \int_0^\infty R_\lambda(x) \cos(\lambda y) d\lambda. \tag{10.7}$$

An arbitrary small contour perturbation can also be represented in the form of the Fourier integral:

$$a(y) - a_0 = \int_0^\infty \beta_\lambda \cos(\lambda y) d\lambda. \tag{10.8}$$

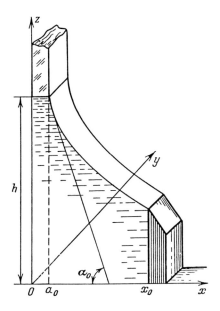

Fig. 10.1. Liquid meniscus parameters: α_0 – the plate half-thickness; x_0 – the shaper half width.

Then the criterion of perturbation infinitesimal will take the following form:

$$\frac{\beta_\lambda}{x-a} \ll 1; \qquad \lambda\beta_\lambda \ll 1 \tag{10.9}$$

where β_λ is the expansion coefficient and $2\pi/\lambda$ is the perturbation wavelength.

Obviously, R_λ linearly depends on β_λ, i.e., $R_\lambda(x) = f_\lambda(x)\beta_\lambda$. The following equation for $f_\lambda(x)$ can be derived:

$$f_\lambda'' - f_\lambda' 6z_0 P(1+P^2)^{1/2} - f_\lambda[\lambda^2(1+P^2) + 2(1+P^2)]^{3/2} = 0 \tag{10.10}$$

with the boundary conditions following from (10.2), (10.3), (10.5) and (10.6):

$$f_\lambda \mid_{x=x_0} = 0 \tag{10.11}$$

$$f_\lambda \mid_{x=a_0} = -P_0. \tag{10.12}$$

To calculate $\mathrm{tg}\alpha_0$ the derivative of z with respect to the normal towards the contour should be found. However, in the linear approximation it is sufficient that the derivative of z with respect to the x-axis should be found, i.e.:

$$\frac{dz}{dx} \mid_{x=a(y)} = -\mathrm{tg}\alpha_0 = \frac{dz_0}{dx} \mid x = a(y) + \frac{dz_1}{dx} \mid x = a(y) =$$

$$= \mathrm{tg}\alpha_e + z_0'' \mid x = a_0[a(y) - a_0] + \frac{dz_1}{dx} \mid_{x=a_0} . \tag{10.13}$$

Then for small difference $\delta\alpha_0 = \alpha_0 - \alpha_e$

$$\delta\alpha_0 = -\int_0^\infty \frac{f_\lambda\mid_{x=a_0} + z_0''\mid_{x=a_0}}{1 + P_0^2}\beta_\lambda\cos(\lambda y)dy. \tag{10.14}$$

The rate of changes in $a(y)$ is determined by (1.38):

$$\frac{d}{dt}[a(y) - a_0] = -V\delta\alpha_0. \tag{10.15}$$

Substitution of (10.8) and (10.14) into (10.15) gives:

$$\int_0^\infty \dot{\beta}_\lambda\cos(\lambda y)dy = V\int_0^\infty \frac{f_\lambda'\mid_{x=a_0} + z_0''\mid_{x=a_0}}{P_0^2}\beta_\lambda\cos(\lambda y)dy. \tag{10.16}$$

where a dot denotes differentiation with respect to time. Thus,

$$\frac{\dot{\beta}_\lambda}{\beta} = V\frac{f_\lambda'\mid_{x=a_0} + z_0''\mid_{x=a_0}}{1 + P_0^2}. \tag{10.17}$$

If $\lambda = 0$ the ratio $\dot{\beta}_0/\beta_0$ represents the above-discussed coefficient $A_{RR} = -V\partial\alpha_0/\partial R = -V\partial\alpha_0/\partial a_0$. Similarly, the equation $\dot{\beta}_\lambda/\beta_\lambda = -V\partial\alpha_0/\partial\beta_\lambda$ holds for all λ values. It will be shown that with λ increase the value of the ratio, $\dot{\beta}_\lambda/\beta_\lambda$ decreases. For this purpose it is sufficient to prove that

$$\frac{\partial}{\partial\lambda}(f_\lambda'\mid_{x=a_0}) < 0. \tag{10.18}$$

Equation (10.10) will be differentiated with respect to λ and the function $F_\lambda = \partial f_\lambda/\partial\lambda$ will be introduced. Then for F_λ the following equation and boundary conditions hold:

$$F_\lambda'' - F_\lambda'6z_0P(1 + P^2)^{1/2} - F_\lambda[\lambda^2(1 + P^2) + 2(1 + P^2)^{3/2}]$$
$$= 2\lambda(1 + P^2)f_\lambda$$
$$F_\lambda\mid_{x=x_0} = 0, \qquad F_\lambda\mid_{x=a_0} = 0. \tag{10.19}$$

From (10.19) it follows that if $f_\lambda > 0$ the function $F_\lambda \leq 0$ since the maximum of this function cannot lie within the range of its positive values. Indeed, if $f_\lambda > 0, F_\lambda > 0$, at the extremum point $F_\lambda' = 0$ and $F_\lambda'' > 0$, which is the condition of its minimum. Besides, $F_\lambda\mid_{x=a_0} = 0$ and since for the rest of the interval $F_\lambda < 0, F_\lambda'\mid_{x=a} < 0$ and (10.18) proves to hold.

Now it will be shown that indeed $f_\lambda > 0$. Since $P_0 < 0$ ($P_0 = -\text{tg}\alpha_e$) from Equation (10.10) and the boundary conditions (10.11) and (10.12) it follows that $f_\lambda > 0, f_\lambda' < 0$. It is attributed to the fact that the maximum of f_λ cannot be found in the range of its positive values, and its minimum cannot lie within the range of its negative values. Hence, f_λ monotonically decreases remaining positive since $f_\lambda = -P_0 > 0$ if $x = a_0$.

Thus, a conclusion can be drawn that if the system is stable in a capillary sense relative to uniform perturbations ($\lambda = 0$, $A_{RR} = \dot\beta_0/\beta_0 < 0$) *it is stable relative to arbitrary perturbation* ($\dot\beta_\lambda/\beta_\lambda < \dot\beta_0/\beta_0 < 0$). It also holds for the wetting boundary condition satisfied on the shaper wall.

As was shown in 3.4 plate growth under the catching condition is stable in a capillary sense relative to uniform perturbations ($\lambda = 0$, $A_{RR} < 0$). Hence, it is stable relative to arbitrary perturbations.

As an example, plate growth under zero pressure ($d = 0$) and the conditions when the melt column weight can be ignored will be considered ($x_0 - 2a_0 \ll 1$). In this case from (10.10) it follows that $dz_0/dx \equiv P = P_0 < 0$ does not depend on x. Integration of (10.10) with the boundary conditions (10.11) and (10.12) and assumption that $P = P_0$ give:

$$\frac{\dot\beta_\lambda}{\beta_\lambda} = -V \mid P_0 \mid \frac{\lambda \operatorname{cth}[\lambda\sqrt{1 + P_0^2}(x - a_0)]}{1 + P_0^2} < 0 \qquad (10.20)$$

for all λ-values.

10.2. Allowance for Morphological Instability of the Crystallization Front

Let us analyze whether the results of the previous paragraph will change when morphological instability of the crystallization front shape is taken into consideration [143].

The overall problem should be formulated in the following way. Interrelation between the distortions of the crystal cross-section shape and those of the crystallization front shape should be allowed for, and for this purpose the Laplace capillary equation should be solved simultaneously with the thermal-conductivity equation and the impurity-diffusion equation. Therefore, its simplified version, i.e., the influence of crystallization front distortions on the crystal cross-section shape, will be considered as the first approximation. Then at first, as has already been mentioned, the case of plate-shaped crystal growth considered in the previous paragraph will be investigated. Where the plate width is infinitely large, similarly to the small deviations of $z_1(x, y)$-values of the perturbed vertical coordinate of the meniscus surface from the unperturbed $z_0(x)$-value, those of the crystal contour $x_1(y) = a(y) - a_0$ and the meniscus shape $h_1(y)$ can be represented in the form of the Fourier integral as in the previous paragraph:

$$z_1(x, y) = \int_{-\infty}^{\infty} R_\lambda(x) \exp(i\lambda y) d\lambda \qquad (10.21)$$

$$x_1(y) = \int\limits_{-\infty}^{\infty} \beta_\lambda \exp(i\lambda y)d\lambda \tag{10.22}$$

$$h_1(y) = \int\limits_{-\infty}^{\infty} \gamma_\lambda(x) \exp(i\lambda y)d\lambda. \tag{10.23}$$

The unperturbed meniscus is described by Equation (10.1) with the following boundary conditions at the crystal and shaper free-edges satisfied (the pressure is assumed to be equal to *zero*)

$$z_0'(a_0) = P_0, \qquad z_0(x_0) = 0 \tag{10.24}$$

respectively, where $P_0 = dz_0(x)/dx$, P_0 usually does not exceed *zero* and x_0 can be any, even infinitely large (the Czochralski technique). The perturbation $z_1(x, y)$ satisfies Equation (10.4) in the same way as in the previous paragraph. The difference is that the crystallization front surface can also be perturbed, consequently the boundary conditions assume the following form:

$$z_1(x_0, y) = 0, \qquad z_1(a_0, y) + P_0 x_1(y) = h_1(y). \tag{10.25}$$

Correspondingly, for Fourier's components the following equalities can be written down:

$$R_\lambda(x_0) = 0, \quad R_\lambda(a_0) = -P_0\beta_\lambda + \gamma_\lambda, \quad R_\lambda(x) = f_\lambda(x)\beta_\lambda \tag{10.26}$$

where $f_\lambda(x)$ is specified by Equation (10.10). If it is assumed that $R_\lambda(x) = S(x)(\gamma_\lambda - \beta_\lambda P_0)$ the equation for the function $S(x)$ is the same as that for $f_\lambda(x)$ and at the boundaries it assumes the following values:

$$S(x_0) = 0, S(a_0) = 1. \tag{10.27}$$

The coefficient of $S(x)$ in (10.10) is negative, therefore the signs of $S''(x)$ and $S(x)$ coincide at the extreme points. It results in such a situation that neither positive maxima (since at the maximum point $S''(x_{max}) \leq 0$) nor negative minima where $S''(x_{min}) \geq 0$ of the function $S(x)$ exist within the interval $(a_0; x_0)$. Thus, $S(x)$ monotonically decreases and its derivative $S'(x)$ does not exceed zero. For α_0-deviations from the equilibrium value the following expression can be obtained in the same way as in the previous paragraph:

$$\delta a_0 = \frac{d^2 z_0(x)/dx^2 \mid_{x=a_0} x_1(y) + d^2 z_1(x)/dx^2 \mid_{x=a_0}}{1 + p_0^2}. \tag{10.28}$$

For deviations $\delta\varphi$ of the azimuthal normal to the contour angle from its equilibrium value the following expression can be obtained:

$$\delta\varphi = -\frac{\delta z_1(x, y)}{dy} \mid_{x=a_0} P_0. \tag{10.29}$$

From (10.15), (10.28) for an isotropic crystal it follows that

$$\dot{\beta}_\lambda = D_{1\lambda}\beta + D_{2\lambda}\gamma_\lambda, \tag{10.30}$$

where

$$D_{1\lambda} = V E'_\lambda(x_0)[\sin^2 \alpha_0(1 + 1/P_0^2)]^{-1} \tag{10.31}$$

$$D_{2\lambda} = V S'_\lambda(x_0)[\sin^2 \alpha_0(1 + 1/P_0^2)]^{-1}. \tag{10.32}$$

$E_\lambda(x) = -S(x)/P(x)$, the latter satisfying the following equation

$$E''_\lambda + 2E'_\lambda P z_0(1 + P^2)^{1/2}(2/P^2 - 1) - E_\lambda \lambda^2(1 + P^2) = 0 \tag{10.33}$$

with the following boundary conditions:

$$E_\lambda(a_0) = P_0^{-1}, \qquad E_\lambda(x_0) = 0. \tag{10.34}$$

If it is assumed that $P_0 < 0$, $\lambda > 0$, an analysis similar to that for $S(x)$ shows that the function $E_\lambda(x)$ monotonically decreases within the range of $[a_0, x_0]$ and the derivative $E'_\lambda(a_0) < 0$, therefore the coefficients $D_{1\lambda}$ and $D_{2\lambda}$ are negative. The second term in (10.30) describes the relation of the m-harmonics of the distortions of the flat front and the radius deviations. From (10.30) it follows that *when morphological stability of the flat crystallization front exists, capillary stability will also be observed. When morphological instability exists, capillary instability will be observed for the same harmonic λ.* The sign of the distortion β_λ is the opposite of the sign of γ_λ since $D_{1\lambda} < 0$; the perturbation decreases with time and the induced one does not. If the meniscus is ambiguous $P(x)$ becomes infinitely large at some point and instead of the function $z(x, y)$ the inverse function $x(z, y)$ is to be analyzed. In this case Equation (10.30) where

$$D_{1\lambda} = V q'_\lambda(h)(1 + g_0^2)^{-1}$$

$$D_{2\lambda} = V [x''_0(h_0) - g_0 q'_\lambda(h_0)](1 + g_0^2)^{-1}$$

also holds for the components β_λ. Here $q_l(h_0)$ denotes the harmonics of the distortion of $x_1(z, y)$, $q_0 \equiv x'_0(z)$.

10.3. Growth of Circular Cylinder-Shaped Crystals by TPS

For axillary symmetric crystals capillary stability of cross-section shapes of crystals of small radii [141] and of arbitrary radii [143] was investigated. The results obtained in [143] that are of more general character will be cited here. As in the previous paragraph, it will be assumed that perturbations extend to the vertical coordinate of the melt meniscus $z(r, \varphi) = z_0 + z_1(r, \varphi)$, to the crystal contour $R(\varphi) = R_0 + R_1(\varphi)$ and to the crystallization front and hence the positions of its points on the crystal contour change, $h(\varphi) = h_0 + h_1(\varphi)$.

From the spacing conditions with respect to the angle φ (Figure 10.2) the function $z_1(r, \varphi)$ should be expanded into a Fourier series with respect to discrete harmonics.

$$z_1(r, \varphi) = \sum_{m=-\infty}^{\infty} u_m(r) \exp(im\varphi) \tag{10.35}$$

and the functions $R_1(\varphi)$ and $h_1(\varphi)$ are expanded in a similar way:

$$R_1(\varphi) = \sum_{m=-\infty}^{\infty} \beta_m \exp(im\varphi); \qquad (\varphi) = \sum_{m=-\infty}^{\infty} \gamma_m \exp(im\varphi). \tag{10.36}$$

An unperturbed meniscus is described by Equation (1.12), while its perturbation $z_1(r, \varphi)$ is described by the following equation:

$$\frac{\partial^2 z_1}{\partial r^2} + \frac{\partial z_1}{\partial r} \left[\frac{1 + 3P^2}{r} - 6z_0 P(1 + P^2)^{1/2} \right]$$

$$+ \frac{\partial^2 z_1}{\partial \varphi^2} \frac{1 + P^2}{r^2} - 2z_1(1 + P^2)^{3/2} = 0. \tag{10.37}$$

As in the previous paragraph, the following equation for β_m can be obtained:

$$\dot{\beta}_m = D_{1m}\beta_m + D_{2m}\gamma_m \tag{10.38}$$

where

$$D_{1m} = W E'_m(r_0)/(1 + P_0^2) \tag{10.39}$$

$$D_{2m} = W S'_m(r_0)/(1 + P_0^2). \tag{10.40}$$

Within the interval of (R, r_0) the behavior of the function $S_m(r)$ is similar to that of $S(x)$ from 10.2, $S'_m(r_0) < 0$. The function $E_m(r) = -S'_m(r_0)/P(r)$ satisfies the following equation

$$E''_m + E'_m M(r) - E_m N_m(r) = 0, \tag{10.41}$$

that can be obtained from the equation for $P(r)$ and $S_m(r)$, i.e., (1.12) and (10.37), where the coefficients $M(r)$ and $N_m(r)$ take the following form:

$$M(r) = (P^2 - 1)r^{-1} + 2z_0(1 + P^2)^{1/2}(2 - P^2)P^{-1} \tag{10.42}$$

$$N(r) = (m^2 - 1)(1 + P^2)r^{-2}. \tag{10.43}$$

The boundary conditions for the functions $E_m(r)$ take the form of (10.34). The case of $m = 0$ (axillary symmetric changes of the crystal diameter) was discussed in Chapter 3. If $m > 1$ the coefficient $N_m(r)$ is positive and according to the result obtained for a plate $E'_m(r_0) < 0$. *In case the conditions of morphological stability are provided and $m > 1$ stability relative to the distortions of the initially cylindrical shape of the crystal cross-section is also observed. However, where morphological instability exists, capillary instability of the crystal shape will be observed.* The results of experiments [142] on growing sapphire filaments by TPS can be explained by this statement. If the growth rate was high enough ($V \sim 0.2$ cm/s) and the filament radius $R \leq 0.01cm$, the surface of the crystallization front represented two bulges (dendrites) towards the melt while the filament cross-section represented an ellipse. Thus, morphological instability led to second-harmonic distortion of the front, which according to (10.38) resulted in distortions of the crystal cross-section shape.

Our experiments on growing sapphire rods with the crystallization front inclined (Figure 3.50) [126] can also be described by the model discussed here. Formally, in this model distortions of the crystallization front shape are introduced by the function $h(r, \varphi)$ alone. In our experiments, this effect was achieved by inclining the crystallization front. One-sided crystal contraction or widening naturally leads to distortions of initially circular crystals.

If $m = 1$ the coefficient $N_1(r) = 0$. The following equation for $P(r)$ can be written down:

$$P'' + P' \left[\frac{1 + 3P^2}{r} - 6z_0 P(1 + P^2)^{1/2} \right]$$

$$-P \left[\frac{1 + P^2}{2} + 2(1 + P^2)^{3/2} \right] = 0. \tag{10.44}$$

It is similar to the equation for $S_1(r)$ that follows from (10.37). For the Czochralski technique $r \to \infty$ and due to similar behaviors of the functions $S_1(r)$ and $| P(r) |$ within the interval of $[R, \infty]$ they are proportional, $E_1(r) = const$ and $E_1'(r) = 0$. So, for the Czochralski technique crystal growth is in indifferent equilibrium relative to transverse displacement of the crystal, i.e., capillary forces do not act at quasi-static transverse displacement. Transverse crystal displacement (with 'bend' formation) can take place only due to the first-harmonic morphological instability, i.e., practically during thermal-center displacement in the process of growth. For TPS it can no longer be affirmed that the functions $S_1(r)$ and $P(r)$ are proportional. The solution of the equation for $E_1(r)$ has the following form:

$$E_1(r) = -E_1^*(R) \int_r^{r_0} \exp[- \int_R^{r*} M(r^{**})dr^{**}]dr^* \tag{10.45}$$

where

$$E_1^*(R) = \left\{ P_0 \int_R^{r_0} \exp[- \int_R^r M(r^*)dr^*]dr \right\}^{-1}. \tag{10.46}$$

From (10.46) it follows that if $P_0 < 0$ and $R < r_0$ then $E_1^*(R) < 0$. Similarly, if $P_0 > 0$ and $R > r_0$ then $E_1^*(R) < 0$. It agrees with the four types of one-valued meniscus shown in Figure 3.20a, b, c, d. For multivalued meniscus (Figure 3.28), the same analysis can be performed using the monotony intervals (r_0, R_0) and (R_0, R_{01}, R_{02}). Since $P(R_0) = \infty$, $E_1^*(R_0) = 0$ and substitution of r_0 by R_0 in (10.46) makes $E_1^*(r_0)$ negative. *Thus, for all the radii of the shapers and the crystals* *being grown the first-harmonic TPS growth will be stable where morphological stability exists.*

Capillary stability relative to meniscus distortions ($m > 1$) and transverse crystal displacements ($m = 1$) means that for a coaxial position of the crystal and the shaper at the specified meniscus height its free-energy minimum composed of the potential and surface energies is achieved. Any small meniscus distortion β_m caused by crystal displacement relative to the shaper (similar to that in the floating-zone technique) or by distortions of its cross-section shape results in an increase of meniscus free energy. Thus, free energy is a quadratic function of β_m. The same is valid for the case when $\mid P_0 \mid = \infty$ (the growth angle $\psi_0 = 0$). *When morphological stability is absent and this leads to the situation when the crystallization front height on the lateral surface of the crystal exhibits a distortion harmonic m the same-harmonic instability of the crystal cross-section shape occurs.* Qualitatively it was proved by experiments in [142] and by our filament-growth experiments described above (Figure 3.50) [126].

Based on determining the free-energy minimum, conservation of cross-section shapes of crystals of complex profiles grown by TPS in each particular case can be analyzed. The overall free energy takes the following form:

$$\Phi = \int\limits_0^h \int\limits_0^{2\pi} \left[\rho g z \frac{r^2(z,\varphi)}{2} + \gamma_{LG} r(z,\varphi) \sqrt{1 + r_z'^2(1 - r_\varphi'^2/r^2)} \right] d\varphi dz \quad (10.47)$$

where a singly connected surface is considered for simplicity. Now, specification of the deviation

$$r_1(z,\varphi) = \sum_{m=\infty}^{\infty} R_m(z) \exp(im\varphi) \quad (10.48)$$

and substitution of $r_0(z,\varphi)$ calculated from the Laplace equation (1.9) where $r(z,\varphi) = r_0(z,\varphi) + r_1(z,\varphi)$ into (10.47) give the change in the free energy $\Phi_1 = \Phi - \Phi_0$. If for all the m harmonics the deviation Φ_1 has the form:

$$\Phi_1 = \int\limits_0^h \left[\sum_{m=-\infty}^{\infty} C_m R_m^2(z) \right] dz \quad (10.49)$$

where $Re(C_m) > 0$, i.e., quadratic relative to $R_m(z)$, the shape of the crystal cross-section is capillary stable. If for the some harmonic $n Re(C_n) < 0$ in (10.49) or the quantity Φ_{1n} has the form:

$$\Phi_{1n} = \int\limits_0^h C_n R_n(z) dz$$

i.e., linear relative to $R_n(z)$, the cross-section shape will be unstable relative to these harmonic distortions.

Chapter 11

SHAPED CRYSTAL GROWTH BY TPS

11.1. Impurity Distribution

The quality of crystals being grown depends to a considerable extent on homogene-ity, of distribution of both specially adder and detrimental impurities. Variations of concentration of such impurities along the crystal length and cross-section are determined by the processes that take place in the melt near the crystallization front.

Impurity distribution in the melt growth of crystals by the Bridgman, Czochralski and floating-zone techniques has been studied extensively, however extension of the mechanisms known to thin-profile growth using capillary fed shapes can lead to inaccurate conclusions.

Application of the BPS equation [144] to calculation of the effective coefficient K of impurity distribution requires specification of the thickness of the boundary diffusion layer. Where this thickness is assumed to be equal to the total height of the meniscus and the shaper capillary channel, $K = 1$ for any impurity [145].

The present chapter gives a model of impurity transfer for the growth scheme under consideration that allows us to relate K-values to the parameters of capillary shaping and feeding [146].

Figure 11.1a illustrates the case of silicon-tape growth. The stationary process is assessed here; it is assumed that the melt in the zone of the meniscus and the capillary channel has not been stirred and the conditions of complete stirring are maintained in the crucible. Under the assumptions made, impurity transfer in the meniscus is described by the equation:

$$D \left(\frac{d^2C}{dr^2} + \frac{1}{r}\frac{dC}{dr} \right) = -VdCdr \tag{11.1}$$

with the following boundary condition at the crystallization front:

$$-D\frac{dC}{dr}\Big|_{r=r_0} = V_0(1 - K_0)C(r_0). \tag{11.2}$$

Here C denotes impurity concentration in the melt, D is the diffusion coefficient,

K_0 is the equilibrium coefficient of impurity distribution, other notations are given in Figure 11.1a.

The polar coordinate system selected allows easy specification of the melt-flow rate distribution in the meniscus $V(r) = V_0 r_0 / r$. Then Equation (11.1) is rearranged in the following form:

$$\frac{d^2C}{dr^2} + \frac{1}{r}\left(1 + \frac{V_0 r_0}{D}\right)\frac{dC}{dr} = 0. \tag{11.3}$$

Impurity transfer in the capillary channel is described by the equation:

$$D\frac{d^2C}{dr^2} = -V_c\frac{dC}{dz} \tag{11.4}$$

with the following boundary condition:

$$C \to \infty, \quad \text{if} \quad z \to \infty \tag{11.5}$$

It is assumed that the impurity concentration in the bulk of the melt C_∞ is constant in the process of tape growth.

The first integral of Equation (11.3) is as follows:

$$\frac{dC}{dr} = A_1 r^{-e-1}$$

where $e = V_0 r_0 / D$. Employment of condition (11.2) gives:

$$\frac{dC}{dr} = -\frac{V_0}{D}(1-K_0)C(r_0)\left(\frac{r_0}{r}\right)^{e+1}$$

$$C(r) = (1-K_0)C(r_0)\left(\frac{r_0}{r}\right)^e + A \tag{11.6}$$

Integration of Equation (11.4) with the boundary conditions (11.5) gives:

$$\frac{dC}{dz} = B_1\exp(-\beta z), \quad \beta = \frac{V_c}{D}$$

$$C(z) = -\frac{B_1}{\beta}\exp(-\beta z) + C_\infty. \tag{11.7}$$

Using the geometrical relation $z = (r^2 - r_c^2/4)^{1/2}$, Equation (11.7) can be rearranged as follows:

$$C(r) = -\frac{B_1}{\beta}\exp\left(-\beta(r^2 - r_c^2/4)^{1/2}\right) + C_\infty$$

$$\frac{dC}{dr} = -\frac{\beta}{(1-r_c^2/4r^2)}\exp\left(-\beta(r^2 - r_c^2/4)^{1/2}\right). \tag{11.8}$$

To define the constants A_2 and B_1 two boundary conditions should be formulated. The first of them follows from equality of concentrations defined from Equations (11.6) and (11.8) at the lower meniscus boundary ($r = r_\phi$):

$$C^m(r_\phi) = C^c(r_\phi). \tag{11.9}$$

Here the indices m and c relate to Equations (11.6) and (11.8), respectively.

The second boundary condition is the condition of impurity-flow balance at the lower meniscus boundary:

$$-D\theta\frac{dC^m}{dr}\Big|_{r=r_\phi} + D_\varphi\frac{dC^c}{dr}\Big|_{r=r_\phi} = \theta\frac{V_0 r_0}{r_\phi}C^m - \varphi V_c C^c\Big|_{r=r_\phi}. \tag{11.10}$$

Employment of these conditions gives an expression describing impurity concentration distribution in the meniscus:

$$C(r) = (1-K)C(r_0)\left(\frac{r_0}{r}\right)^e$$
$$+ \frac{C_\infty - (1-K_0)(r_0/r_\phi)^e(\cos\varphi - 1)C(r_0)}{1 + [(V_0\theta r_0/V_c\varphi r_\phi) - 1]\cos\varphi}. \tag{11.11}$$

The notations $\cos\varphi = (1 - r_c^2/4r_\phi^2)^{1/2}$, $e = V_0 R/D\sin\theta$ are used here; if Equation (11.11) is written down for the point $r = r_0$ and the expression of the equilibrium $K_0 = C_S/C(r_0)$ and the effective $K = C_S/C_\infty$ coefficients of distribution (C_S is the solid-phase impurity distribution) are used, an expression for K can be derived:

$$K = \frac{K_0}{K_0[1 + (V_0\theta r_0/V_c\varphi r_\phi - 1)\cos\varphi] + (1-K_0)(1 - \cos\varphi)(r_0/r_\phi)^e}. \tag{11.12}$$

With the following simple geometrical relations applied: $2V_0 R = V_C r_C$, $r_0/r_\phi = 2R/t_\phi$, $r_0 = R/\sin\theta$, Equation (11.12) assumes the form:

$$K = \frac{K_0}{K_0[1 + (V_0 r_C/\varphi t_\phi - 1)\cos\varphi] + (1-K_0)(1 - \cos\varphi)(2R/t_\phi)^{V_0 R/D\sin\theta}}. \tag{11.13}$$

In case the width of the capillary channel reaches its maximum value $r_C = t_\phi$, Equation (11.13) is reduced to:

$$K = \frac{K_0}{K_0 + (1-K_0)(1 - \cos\varphi)(2R/t_\phi)^{V_0 R/D\sin\theta}}. \tag{11.14}$$

Equation (11.13) relates the value of the effective coefficient of impurity distribution K with the parameters of the crystallization conditions (V_0 and $2R$), with the conditions of capillary shaping (t_ϕ, r_C, θ) and the impurity characteristics (D and K_0).

Figure 11.1b shows $K = f(r_C)$ calculated for aluminum impurities in silicon. The following values are used: $K_0 = 0.002$, $D = 5.3\ 10^4 \text{cm}^2\text{s}^{-1}$, $2R = 0.03$ cm, $t_\phi = 0.06$ cm.

Figure 11.1c shows $K = f(V_0)$ calculated from Equation (11.13) for various r_C values.

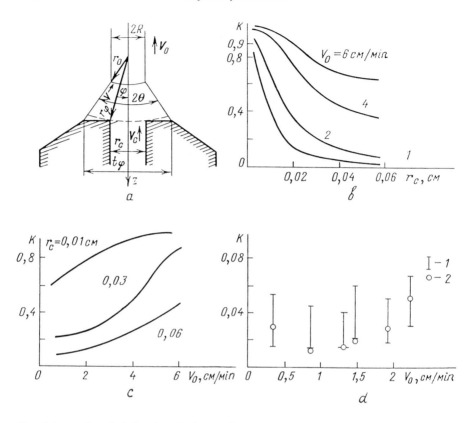

Fig. 11.1.   On calculating the effective coefficient of impurity distribution: (a) diagram; (b) $K = f(r_c)$-calculated; (c) $K = f(V_0)$-calculated; (d) comparison of experimental (1) and calculated from (3.20) (2) values of the effective coefficient of In distribution for thin-walled profiled Si crystal growth.

It is obvious that the effective coefficient of impurity distribution increases as the growth rate and the melt flow rate in the shaper capillary channel increase. Under actual conditions of growing thin-walled profiled silicon crystals $K_{Al} \ll 1$.

Experimental data allowing thorough analysis of the above-derived relations are not currently available. Processing of the data of [147] gives the value of $K = 0.039$ for aluminum impurity in silicon, which agrees very well with the value of $K = 0.3$–0.4 calculated from Equation (11.13) in accordance with the initial data of [147]. The results obtained using the mass spectrograph analysis of impurity content in tapes grown from refined metallurgical silicon are given in Table 11.1. They confirm that the effective coefficient of distribution of the impurities analyzed is much less than *one*.

Figure 11.2 gives data on sulphur distribution along the axis of a silicon tape obtained by laser emission microanalysis (LEM). The ratio of the sulphur spectral

Table 11.1.

material	impurity, 10^4 mass %					
	Cu	Mg	B	Ni	Mn	Fe
silicon precursor	1	0.5	1	0.1	1	5
silicon tape	1	5	1	1	1	1

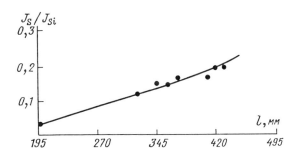

Fig. 11.2. Sulphur distribution along the length of a silicon tape (LEM-data).

line strength to that of silicon defined from the lines of Si ($\lambda = 288.1$ nm, $\lambda = 363.2$ nm) is plotted on the ordinate (the J_{el}/J_{Si} ratio is proportional to the concentration of the element analyzed in the silicon matrix). Increase of sulphur concentration in the silicon tape in the process of its pulling shows that $K_S \ll 1$. (It is assumed that for sulphur impurity in silicon $K_0 = 110^{-5}$). The rate of tape pulling was equal to 12 mm/min. While analyzing other impurities, no explicit regularities were observed.

To verify the main Equation (11.13) a series of 0.01 mass indium-doped silicon tapes was grown.

The mass of each tape-shaped crystal did not exceed 7% of that of silicon charged into the crucible. The tape width $2R$, the shaper transverse dimension t_Φ, the capillary slot dimension r_c, the growth rate V_0 were measured. Then photometrical spectral lines of indium in the crystals grown and in the crucible residue were drawn. The measurement accuracy of the effective coefficient of indium distribution was equal to (40-50)%.

The results obtained are depicted in Figure 11.1d. Vertical arrows indicate experimental results, circles show the values calculated from Equation (11.13) in accordance with the above-mentioned parameters. The values of $K_0 = 4 \cdot 10^{-4}$, $D = 5.2 \cdot 10^{-4}$cm^{2}s^{-1} are assumed for indium impurity in silicon.

The results shown in Figure 11.1d do not allow K to be plotted versus the growth rate V_0 since the thickness of a silicon tape $2R$ decreases as V_0 increases (the value of $2V_0R$ hardly changed in the experiments), and they can only demonstrate satisfactory agreement between calculated and experimental values. The values of K calculated from the Burton-Prim-Slichter equation are equal to 0.8–0.9, i.e., they are some

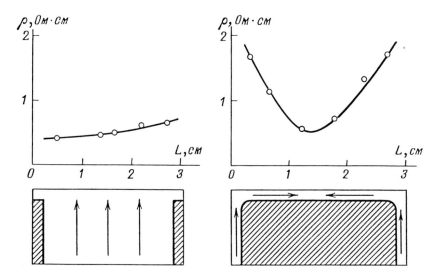

Fig. 11.3. Resistivity distribution along the widths of silicon tapes grown using two various versions of liquid-phase replenishment of the meniscus.

orders of magnitude greater than the experimental values.

Impurity distribution along the width of the profiled crystals is to a great extent determined by the technique used to feed the melt to the growth meniscus. In Figure 11.3 resistivity distributions ρ along the width of silicon tapes grown under the conditions of the two versions of meniscus melt replenishment are compared. The shaper shown in Figure 11.3a possesses one long capillary slot, while the shaper given in Figure 11.3b has two short slots at its end faces.

The following explanation of non-uniformity of impurity distribution along the crystal width (Figure 11.3b) observed can be offered. In the case of horizontal melt flow in the meniscus from the shaper edges towards its center, impurities with $K < 1$ driven off by the growing crystal accumulate in the central part of the meniscus. Hence, corresponding distribution of capillary channels in the shaper allows control of impurity distribution along the width of crystals being grown.

11.2. The Growth of Shaped Silicon Crystals by TPS

11.2.1. Silicon-Tape Growth Procedure

Silicon tapes were grown using equipment with both resistance furnaces and induction heating [148]. In the former case, the heat zone was made to the shape of the rectangle so that it could follow the shape of the pulled crystal cross-section as closely as possible. This heat zone was used to grow silicon tapes up to 40 mm wide. The

Table 11.2.

parameter	meniscus flat section	meniscus
A_{RR}, s^{-1}	-0.4	-0.10
A_{Rh}, s^{-1}	-0.7	-0.19
A_{hR}, s^{-1}	3.37	2.17
A_{hh}, s^{-1}	-0.46	-0.42
$A_{RR} + A_{hh}, s^{-1}$	-0.51	-0.51
τ_{rf}, s	4.0	3.8

heat zone of the induction heater was circular in shape. This zone was used to grow silicon tapes 25 mm wide.

Tapes were grown in argon flow in both methods. Six percent of hydrogen added to argon used in the equipment with induction heating helped to avoid break-downs.

Stability analysis. Table 11.2 gives the coefficients A_{ik} characterizing stability and the combinations $A_{RR} + A_{hh}$ and $A_{RR}A_{hh} - A_{Rh}A_{hR}$ for meniscus end faces and midparts where a silicon tape of 0.6 mm is grown at the rate of 2 cm/min. It can be seen that the conditions of process self-stabilization hold, the relaxation time for the perturbations δR and δh is equal to 4 sec. Thus, stability exists and practical problems of thin-walled silicon tape grown using a shaper are caused by small meniscus height and by chemical interaction between the melt, the shaper and the environment.

To stabilize the width of the tape being pulled, its ends could be blown with gas flows through special nozzles. Independent flow control allows effective control of tape widening to the width specified.

The rate of silicon-tape pulling usually reaches (10–15) mm/min unless special coolers are used. Local tape cooling along its perimeter in the vicinity of the crystallization front was used to increase the pulling rate.

11.2.2. Silicon-Tube Growth

A circular heat zone with resistance heating was used to grow silicon tubes [149]. They were grown in argon flow with a rate of up to 6 cm/min. While growing silicon tubes by the semi-continuous technique, in the course of its growth the crystal is withdrawn into the air through a special outlet device preventing ingress of air into the chamber. A contactless water-cooled refrigerator can serve as an outlet device, which allows an additional increase of process output. Now the semi-continuous technique is used to grow silicon tubes with a diameter of up to 12 mm.

Depending on the fields of their application, silicon profiles of various wall thicknesses are required. The wall thickness of a growing tube depends on the shaper design and on the process parameters. However, there are principal restrictions that enable growth of crystals with a specified wall thickness using rates not exceeding

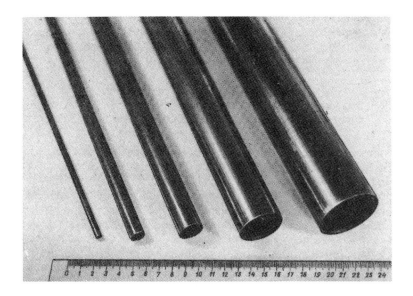

Fig. 11.4.   Specimens of silicon tubes grown TPS.

the maximum growth rate. For example, a tube with a wall thickness of 0.3 mm could be grown with the growth rate of up to 32 mm/min. Tubes with diameters of 50 mm had walls 3 mm thick in certain cases.

Figure 11.4 shows a photograph of silicon tubes grown using a capillary shaper.

Overall stability of the process of silicon-tube growth is so high that several tubes can simultaneously grow from shapers disposed in the crucible. Seeds are also fixed independently in accordance with the shaper arrangement.

Maintaining the same temperature over the shaper perimeter is an important problem of group growth since the operating free edges of the shapers are arranged symmetrically relative to the heater though their axes lie at the corners of the equilateral triangle equally spaced from the axis of the heat zone. For example, when tubes with diameters of 6 mm were grown the wall thicknesses varied within the range of (0.35–0.45) mm over their perimeters. However, proper selection of heat screens allows uniform temperature distribution at the operating free edges of the shapers.

Figure 11.5 shows a photograph of the group-growth process of three silicon tubes, each 6 mm in diameter.

11.2.3. *Growth of Thin-Walled Polyhedral Prism-Shaped Crystals*

A thin-walled polyhedron-shaped crystal can be considered as a number of tapes (depending on the number of its faces) connected with each other. The procedure of growing such crystals differs little from that of growing tubes, therefore the specific feature of the given process is its higher stability as compared with tape growth.

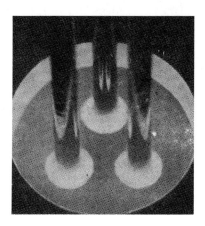

Fig. 11.5.   Group growth of silicon tubes.

To reduce the angle influence on shaping it is necessary to try to increase the number of faces so that two adjacent faces intersect at more obtuse angles. The profiles of square and hexagon-shaped cross-sections were grown [150]. In both cases, the width of each face was equal to 20 mm. Figure 11.6 shows a photograph of hexahedrons and shapers used to grow these crystals.

11.2.4. *Profiled-Silicon Structure*

Profiled silicon crystals possess characteristic defects of their crystalline structures including flat boundaries (most often those of the twinning type), dislocations and pileups thereof as well as SiC particles and particle-aggregations. As a result of twinning along intersecting planes, which can be observed at the initial stage of growth, a stable and quasi-equilibrium structure characterized by the existence of defect areas (flat boundaries) perpendicular to the tape plane and parallel crystal [150–152]. In this case, the silicon tape surface orientation is $\{110\}$ and the direction of pulling coincides with the axis $< 211 >$. Such an orientation is formed and maintained irrespective of the seed orientation and can be attained directly at the seed-crystal contact point when the seed orientation (110) $< 211 >$ or a stable-structure section of the silicon tape is used as a seed. Figure 11.7 shows a photograph of the surface of a silicon tape of a stable defect structure: SiC inclusion surrounded

Shaped Crystal Growth

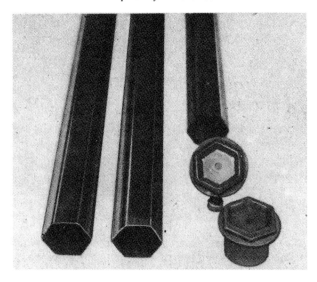

Fig. 11.6. Hexahedral prism-shaped silicon profiles and shapers to grow them.

Fig. 11.7. Defect structure of a silicon tape. SiC-inclusion can be observed in its center (a pattern of selective chemical etching, × 120, H.p. – the direction of crystal growth.

with dislocation pile-ups can be observed on the tape surface.

On the basis of X-ray structure analysis it was established that the orientation of silicon-tape surfaces can vary within the range of 15^0 from $\{110\}$. Considerable dimensional fragments of defect structures identical to those of tapes are observed in silicon tubes (Figure 11.8a). The tube-surface orientation in the vicinity of such a fragment varies from $\{110\}$ due to surface curvature. This change in orientation does not cause additional structure defects up to the values equal to 15^0 and as soon as that value reaches 15^0 surface orientation abruptly changes because there appears 'a severely defect spot' that provides an indispensable turn of the system of twin-boundary planes in such a way that the boundaries are again approximately

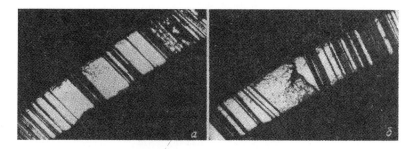

Fig. 11.8. Defect structures of silicon tubes (cross-sectional fragments): (a) typical structure; (b) defect area providing rotation of the system of twin-boundary planes. ($\times$ 200, pattern of selective chemical etching).

perpendicular to the tube surface. Either grain boundaries of a general type or dislocation pileups can act as such defect areas (Figure 11.8b)

Apparently, the number of 'severely defect spots' and correspondingly the number of fragments of steady structures is determined by the necessity of closing the system of fragments into a cylinder. Since the angular magnitude of an arc of 30^0 corresponds to the variation of the tube surface orientation of 15^0 from {110}, the overall number of defect spots is equal to at least twelve, which has been proven experimentally. It should also be noted that this linear length of 'severely defect spots' along the perimeter of a tube depends little on its diameter, therefore the relation of the volume of 'severely defect spots' to crystal volume should quickly decrease as the tube diameter increases.

Defect structures of silicon tapes were studied by the TEM [153]. With $a \sim 200$ magnification, the defect structure pattern proved to be similar to that of selective etching (Figure 11.7). It was observed that the width of defective areas is equal to (3–5) μm and the width of monocrystalline regions between them varies from 10 to 500 μm. Interpretation of fine structures of defect areas, that look like dark lines parallel to the direction of pulling in the pattern of selective chemical etching (Figure 11.7) revealed that each area represents a set of microtwins (40-200)Å wide (Figure 11.9). Small-angle disorientation of monocrystalline sections adjacent to the defect area is caused by a set of microtwins. Disorientation measured by the kikuchi-line technique varied from 40^0 to 4^0. Besides defect, areas of twinning nature, multilayer lattice defects, small-angle dislocation boundaries and separate dislocations and inclusions were observed in 2% of cases.

To interpret the defect structure and macroscopic pattern of profiled silicon crystal growth, a model according to which the crystallization front tends to be shaped by the most slow-growing crystal faces {111} while high rates are provided by availability of inlet angles with their vertexes coinciding with twin boundaries was offered in [154] (Figure 11.10a).

To check the model offered, experiments on investigation of the crystallization front shape visualized by impulse changes in the crystal-pulling rate were carried out.

Fig. 11.9.   TEM image of a silicon tape: flat defect area structure in the form of a system of microtwins ($\times$ 160,000; V denotes the direction of crystal growth).

In this case local changes in crystal thickness follow the changes in crystallization front shape, so does the horizontal hatching observed on the surface of profiled silicon crystals (Figure 11.10d).

Crystallization front shaping by faces $\{111\}$ proved to be observed only in the vicinity of high-energy boundaries of a general type and at twin boundaries of higher orders, e.g., $\{111\} - \{115\}$. Crystallization front deflection from the flat one does not exceed 2 μm in the coherent-twin region. The overall crystallization front area made by faces $\{111\}$ is not large; the major part of the crystallization front follows the crystallization isotherm and corresponds to the face close to (112) and in separate cases to (110) (Figure 11.10) each of which can grow according to the normal mechanism, i.e., both layer-by-layer and normal mechanisms of growth take place when profiled crystals grow.

The specific features of the crystallization front reviled might be associated with the impurity influence on the face free energy or on the growth kinetics. Therefore it can be assumed that front shaping and perhaps inlet angle formation at twin boundaries are not necessary for high rates of profiled silicon crystal growth to be put into practice. It can be confirmed by the fact that in separate cases, silicon tapes with wide (up to 10–15 mm) monocrystalline areas can be grown.

It was noticed that in the process of silicon tube growth the melt meniscus height strongly influences the character of defect formation in crystals. With the crystallization front in a tube sinking, the number of boundaries of a general type increases, i.e., profile defect structure depends on the interaction efficiency in the crystallization front-shaper system. The shaper influence is mainly determined by the distortions

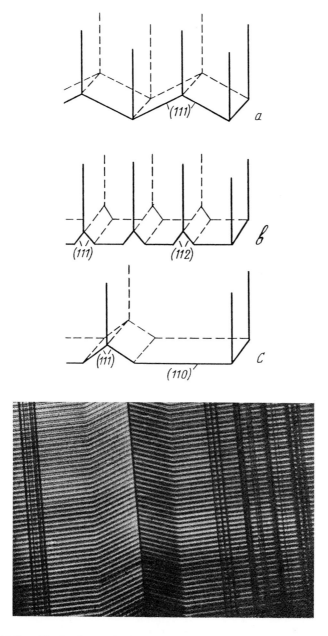

Fig. 11.10. (a) Crystallization front cutting by planes (111) in accordance with the model of [153], (b, c) experimentally observed crystallization front shape; (d) agreement between the crystallization front shape and the horizontal hatching on the surface of profiled crystals.

(thermal and capillary) introduced by the silicon carbide layer formed on the operating free edges of a graphite shaper when it interacts with the silicon melt.

Besides such indirect influence on the structure of profiles, silicon carbide entering the crystal subsurface layer in the form of SiC particles greatly affects the quality of the material produced.

SiC microcrystals measuring (25–30) μm reduced in the melt meniscus form a carbide layer growing on the shaper free edges. When bulges appear on the layer the meniscus becomes distorted and distinctive furrows appear on the crystal surface. In separate cases, bulges along the direction of pulling that break when SiC particles and aggregations result from random fluctuations of the crystallization front when particles reaching the size of the meniscus escape the carbide layer [155]. The number and size of simultaneously entrapped particles depends on the amplitude of front fluctuations and on the mean height of the meniscus.

The mechanism of carbide inclusion entrapment offered is confirmed by the results of silicon tube growth when the number of SiC particles entrapped sharply decreases for a high meniscus since the number of particles whose dimensions reach the meniscus height decreases.

The number of carbide inclusions in profiled silicon crystals can be decreased in several ways: by using crucibles made of graphite of high density; by replacing the graphite container for crucibles with a protective layer-coated container; or by maintaining a high melt meniscus (easy to secure when growing closed profiles). Furthermore, experiments on alloying profiled silicon with rare-earth elements showed that no SiC inclusions can be found on profile surfaces for any meniscus height in case at least 0.05 mass % of gadolinium dopant is introduced into silicon melt.

11.2.5. *Local Electronic Properties of Shaped Silicon*

It is well known that inhomogeneity of the properties stems from the inhomogeneous distribution of electrically active defects. The processes occurring at the solid-liquid $(S-L)$ interface essentially affect formation of defects and their electronic properties as well. The simultaneous study of electric and photoelectric properties of a crystal, its defect structure and the $S-L$ interface was carried out in [150-153]. The electric and photoelectric properties of crystals have been studied by local methods at $T = 300K$: electron-beam induced current (EBIC), spreading-resistance (SR), light-beam induced voltage (LBIV) and light-beam induced current (LBIC).

The EBIC-method is the most widely used. In the SR method, the local conductivity of the sample was detected by an acicular probe displaced on the crystal surface with 8 μm step. In this case the spatial resolution of the method was 30 μm.

The LBIV method was used for measuring the potential difference at the edges of a sample locally illuminated by a scanning light beam of 5 μm in diameter. This method made it possible to reveal the built-in electric fields caused by the defects of the crystal structure.

In order to apply the LBIC method, a $p-n$ junction was prepared or tin and indium oxide film was grown on the silicon surface, and then the current of minority carriers generated by a method analogous to the LBIV method was recorded. So, effective centers of recombination were revealed by the LBIC method.

The spatial distribution of crystal defects was analyzed by the method of selective chemical etching. In addition, the structure of boundaries was examined with a scanning electron microscope.

The morphology of the $S - L$ interface was investigated as was shown above (Figure 11.10d).

EBIC investigations of shaped crystals show that the electrical activity of twin boundaries as centers of increased recombination of nonequilibrium charge carriers differs (Figure 11.11): there are electrically inactive boundaries, which are evidently coherent, boundaries with discontinuous activity, and boundaries of moderate and high activity. In the latter case, the boundaries are evidently incoherent.

A greater emphasis was placed on studding defects whose electrical activity, as recombination centers of nonequilibrium carriers, was constant all along the silicon ribbon. It can be seen from comparing the LBIC spectra (Figure 11.12e) with the pattern of selective chemical etching (Figure 11.12b), that there are at least three types of such defects. They are narrow bands of submicron defects, accumulations or agglomerations of point defects with the coordinate $l = 0.8$–1.2 mm extended along the crystal growth direction. Dislocation etch pits were not visible here. The rows of dislocations denoted by index E in Figure 11.12a and some twin boundaries are highly electrically active.

11.2.6. *Conclusion*

The main purpose of development of shaped silicon growth technology is solar element production. Shaped silicon, at this stage of technology, is characterized by the presence of a defect structure influencing the electronic properties of the crystal. Its influence on the lifetime of secondary charge carriers is of particular importance, since the efficiency of solar elements is determined primarily by the lifetime of the secondary carriers. The investigation of shaped crystals shows the presence of inclusions of SiC particles, block boundaries, acute-angle boundaries, monocrystalline sections with reduced lifetime, incoherent twin boundaries, dislocations, and coherent twin boundaries. The electrical activity of the structural defects falls in the order listed: in the region of an SiC inclusion, the lifetime of secondary carriers $\tau \leq 10^{-8}$ sec, increasing to a magnitude of the order of 10^{-6} sec in monocrystalline regions of Si crystals with a resistivity of the order of $1\,\Omega$cm.

Experiments show that twin boundaries are the main defects. The concentration of electrically active defects may possibly be reduced at the stage of crystal growth by improving the procedure. Decreasing electrical activity of crystal defects in postgrowth treatment is a promising method of increasing the efficiency of solar elements. It has been established that annealing crystals at a temperature of the order of 500°C is largely ineffective. In the EBIC spectra shown in Figure 11.13, the regions of increased recombination of nonequilibrium carriers in the vicinity of the boundaries are seen to broaden only slightly. However, annealing at 1000°C leads to

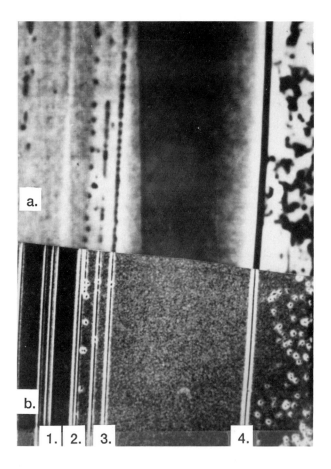

Fig. 11.11. The patterns (a) of spatial distribution of the secondary carrier current (EBIC – the dark regions correspond to increased recombination) and (b) selective chemical etching: recombination boundaries – (1) inactive; (2) intermittent active; (3) low active; (4) high active; and (5) active dislocations.

an increase in lifetime of the secondary carriers as a result of the decrease in electrical activity of the boundaries.

11.3. TPS Growth of LiNbO$_3$ Crystals

11.3.1. *Introduction*

Based on [156, 157], the growth of shaped LiNbO$_3$ crystals of various profiles including multiple growth of ribbon crystals is reported. The techniques for obtaining regular and periodic domain structures in these crystals are studied.

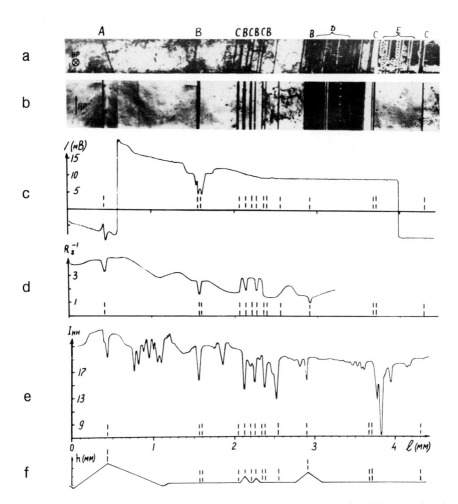

Fig. 11.12. Local study of the defect structure, electric properties and of the $S - L$ interface shape in a silicon ribbon: (a) defect structure of the transverse cross-section of an etched silicon ribbon; (b) structural defects on the surface of an etched ribbon (G.D. indicates growth direction); (c) distribution of photoEMF obtained by the LBIV method across the ribbon; (d) spatial distribution of crystal conductivity, obtained by the spreading resistance method (relative units); (e) spatial distribution of minority carriers across the ribbon (the LBIC method); (f) shape of the $S - L$ interface crystal is shown above the curve, and the melt is below; at $l \geq 2$mm the height l of the $S - L$ interface distortion is expanded by a factor of 10. The dashed lines in (c) to (f) show location of some twin boundaries in (b).

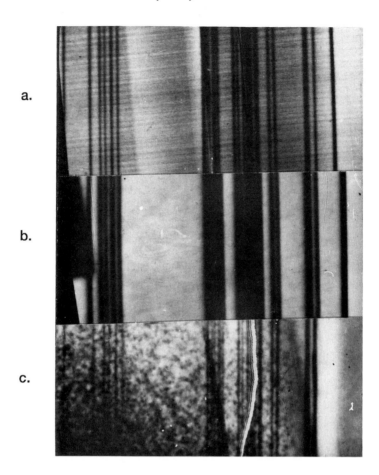

a.

b.

c.

Fig. 11.13. Influence of annealing on the EBIC contrast of silicon strip: (a) initial sample; (b) annealing at 450° for 50hr; (c) annealing at 1000° for 50hr ($\times$ 130).

11.3.2. Technique Description for Growing Shaped LiNbO₃ Crystals

Shaped LiNbO$_3$ crystals were grown in a modified Czochralski crystal grower provided with an induction heater (8000 Hz). The changes introduced were due to the necessity of stabilizing the temperature conditions in the shaper. A special thermal zone was created for growing shaped LiNbO$_3$ crystals (Figure 11.14). An important feature of this zone is that the shaper can be adjusted with respect to the seed crystal, thus enabling one to grow crystals of the desired orientation. The crystals were grown in air from platinum crucibles of various dimensions and shapes. The crucible wall and bottom thickness was 3 mm. The use of thick-walled crucibles appreciably reduces their deformation and increases the stability of the thermal zone. The shapers

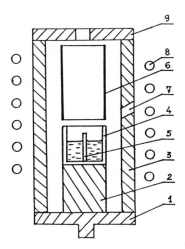

Fig. 11.14. Thermal zone for growing shaped lithium niobate crystals: (1) table, (2) ceramic support, (3) ceramic tube, (4) platinum crucible, (5) platinum shaper, (6) platinum shield, (7) sapphire window, (8) inductor, (9) ceramic cover.

for growing the shaped LiNbO$_3$ crystals were also manufactured from platinum and placed directly into the crucible.

LiNbO$_3$ powder of a congruent-melt composition or Czochralski-grown crystal scrap was used as a charge. Rectangular seeds were cut from precisely oriented lithium niobate crystals grown by the Czochralski technique. The growth rate of the shaped crystals was 5 to 60 mm/h.

In order to reduce thermoelastic stresses that result from growth and cooling, a platinum shield was placed in the thermal zone. This shield was heated by the same inductor as the crucible. In addition, passive shields of sintered aluminum oxide tubes were used. The crystallization process at the shaper end surface was monitored by means of a long focal length microscope through the small openings in the tubes. These openings in the thermal zone were later closed by a sapphire window because of undesirable convective gas flows.

The crystals were cooled directly in the thermal zone in two stages in the following way: (1) From the melting point to 500°C, the temperature in the zone was decreased linearly at a rate of 50 to 250^{0}C/h by means of a programmed controller. (2) The final cooling from 500°C down to room temperature was obtained by turning off the *RF* generation and the cooling water. The overall cooling process took 6 to 12 h depending on the crystal dimensions.

The above technique was employed to obtain ribbons, tubes and other profiles up to 150 mm long (Figure 11.15a). A simultaneous growth of several LiNbO$_3$ ribbons was successfully performed (Figure 11.15b).

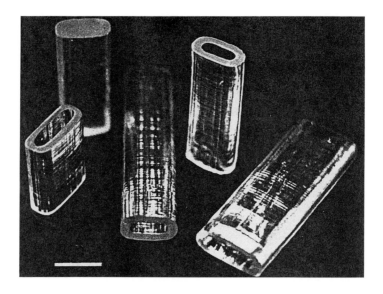

Fig. 11.15.   Samples of shaped lithium niobate crystals (a); lithium niobate ribbon crystals grown simultaneously (b). Markers represent 1 cm.

11.3.3. *Control of Domain Sizes in Shaped Lithium Niobate Crystals*

The presence of a spontaneous dipole moment causes the formation of the domain structure of non-polarized ferroelectric $LiNbO_3$ crystals. Because the point group

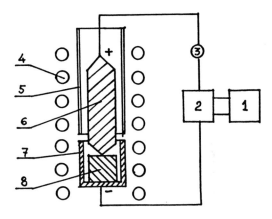

Fig. 11.16. Scheme of polarization of shaped LiNbO₃ crystals: (1) power supply, (2) current stabilizer, (3) millimeter, (4) inductor, (5) platinum shield, (6) crystal, (7) platinum crucible, (8) shaper.

symmetry is 3m with a special direction C, the domains are 180°. There are several techniques to obtain monodomain LiNbO₃ crystals. The Czochralski grown LiNbO₃ crystals are normally polarized in special annealing furnaces. In contrast shaped lithium niobate crystals were polarized directly in the process of growth.

The polarization technique for shaped LiNbO₃ crystals is demonstrated in Figure 11.16. The system for obtaining a single domain was turned on immediately after the crystal was enlarged to the desirable dimensions. At the end of the process, in order to prevent dielectric breakdown, the crystal neck was reduced to 2 to 3 mm in cross-section, after which pulling was stopped. The polarization system was turned off as the die temperature was decreased below the Curie temperature. $T_C = 100°C$. The magnitude of the current density was 1 - 2.5 mA/cm², the polarity being (+) for the seed and (-) for the crucible.

It is known that if in the process of the Czochralski growth of LiNbO₃ the direction of the applied electrical field is reversed, then the 180° domains are formed. Analogous experiments were carried out during growth of shaped LiNbO₃ crystals. It should be noted that, in this case, the sign-variable electric field can be applied to the crystal not only in the growth process but during cooling as well; that is, one can control the domain structure in the part of the crystal which has a temperature above T_c immediately after pulling is stopped. By varying the growth rate, cooling rate and the period of the applied sign-variable electric field, domains from several cm to several tens of μm and smaller in size were obtained. Figure 11.17 demonstrates one of the domain structures obtained.

The shape of the domains is related to the Curie isotherm, therefore it is possible to select optimal thermal conditions of the crystallization process. This problem will be treated separately elsewhere.

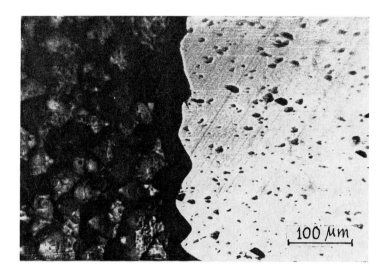

Fig. 11.17. Domain structure formed in a sign-variable electric field during the crystal cooling.

11.3.4. *Regular Domain Structure in Growth from Doped Melts*

A periodic domain structure can be formed in lithium niobate crystals grown by the Czochralski technique from the melt with scandium, indium and yttrium as dopants. This structure repeats the crystallization front shape and is attributed to periodic distribution of the impurity in the crystal. In order to change the domain dimensions and their periodicity, an asymmetric thermal field is usually used. Under these growth conditions, periodic temperature fluctuations of the melt occur at the crystallization front, leading to a change of the crystallization rate and, consequently, to a periodic incorporation of the impurity into the crystal. The frequency of the temperature fluctuations can be varied by varying the rotation and pulling rates.

Shaped lithium niobate ribbon crystals were grown by the above-described technique. The crystals were doped with 1 wt% $LiTaO_3$ [158]. The growth direction coincided with the [0001] direction, the $(10\bar{1}0)$ plane being the ribbon plane.

The impurity distribution in the crystals and hence their domain structure was controlled by varying the pulling rate rather than by producing a periodic temperature fluctuation in the melt under the crystallization front. This is associated with the fact that due to thermal inertia of the die it is very difficult to get a desired frequency of temperature fluctuations. The pulling rate was varied periodically and discontinuously. Under these growth conditions, there occurs a periodic alteration of the shaped crystal cross-section, which leads to a change of the effective coefficient of the impurity distribution. Thus, domains measuring 200 to 50 μm were obtained (Figure 11.18). In as much as the shape of the domains reproduced that on the shaper end surfaces, we obtained horizontal, spherical, conical as well as some other shapes

Fig. 11.18. Domain structure formed at discontinuous alteration of the pulling rate.

of the domain structure by varying the shaper geometry.

11.4. The Growth of Shaped Sapphire Crystals by TPS

Sapphire single crystals formed as tubes with a circular cross-section seem rather appropriate for illumination applications. The properties of sapphire, such as its high melting point, high translucence, strength, and resistance to alkali-metal vapors at high temperatures, enable the use of this material as envelopes in high pressure gas-discharge lamps which use a discharge in alkali-metal vapors.

There is some experience in manufacturing high pressure sodium lamps with polycrystalline aluminum (Lucalox) envelopes. The luminosity of such lamps is rather high $(> 100 lmW^{-1})$. The use of single crystal sapphire tubes for high pressure lamps makes their luminosity even higher and increases their life. The experience gained from utilization of high pressure sodium lamps with single crystal sapphire tubes has shown that their life is twice as long as that of lamps with Lucalox envelopes. Due to their low power consumption and good spectral characteristics, high pressure sodium lamps are the most promising general purpose light source.

The main aim of this section is to describe the most significant characteristics of the defect structure of sapphire tubes manufactured by TPS. We used tube samples which enabled us to characterize most clearly the defects formed during the growth process. In the experiments which were carried out, the growth of single sapphire tubes with diameters of 4 to 40 mm and wall thicknesses in the range 0.5 to 3 mm was investigated [159].

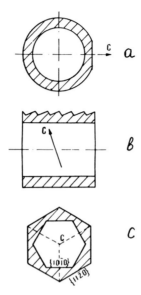

Fig. 11.19.   Faceting of the lateral surface of sapphire tubes grown in different orientations: (a) the cross-section of a tube grown with $\rho = 90°$; the (0001) face is parallel to the pulling direction. (b) The longitudinal section of a tube grown with $\rho < 90°$; a step-like faceting occurs by the (0001) face. (c) The cross-section of a tube grown with $\rho = 0°$; the faceting on the outside occurs by the $\{11\bar{1}0\}$ faces and on the inside by the $\{10\bar{1}0\}$ faces.

11.4.1. *Morphological Features of Sapphire Tubes*

The geometrical form of a tube can differ slightly from the ideal for reasons of a crystallographic nature. According to the Curie theorem, in the process of growth, there is an interaction between the crystal symmetry and that of the medium (i.e., the thermal environment) in which the crystal grows. In the grown crystal, only those elements of symmetry are exhibited which are common both for the crystal and for the medium. That is the reason why grown crystals are faceted depending on the orientation. If a sapphire tube is pulled in the directions $< 10\bar{1}0 >, < 11\bar{2}0 >$ or intermediate ones with $\rho = 90°$ (where ρ is the angle between the c-axis and the pulling direction), the close-packed basal plane (0001) becomes parallel to the growth direction and facets the tube (Figure 11.19a). If ρ differs slightly from 90°, a step-like faceting by the basal plane appears on the lateral surface of the tube (Figure 11.19b). Tubes grown in the [0001] direction are faceted on the outside by $\{11\bar{2}0\}$ planes and on the inside by $\{10\bar{1}0\}$ planes (Figure 11.19c). Faceting is reduced if the temperature gradient at the crystallization front increases.

11.4.2. *Voids*

Voids are one of the major defects in shaped sapphire crystals. Among the causes producing voids in sapphire grown from the melt, one can note the thermal dissociation of the melt, the liberation of dilute gases, and the shrinkage of the melt during crystallization. Although the problem is not completely understood at this time, the thermal dissociation of the melt in contact with the crucible material seems to be the most probable cause of the voids.

Void distribution in the crystal is not uniform. At low crystallization rates (<1 mm/min^{-1}) the voids are absent. The void distribution is at times in the shape of a tree-like pile-up with a branch diameter of up to 1 mm. Figure 11.20a shows an example of this tree-like void distribution in a tube. The distribution of voids over the tube section normal to the growth direction is shown in Figure 11.20b. The characteristic feature of the void distribution over the cross-section at growth rates of 1 to 3 mm/min^{-1} is the presence of a 50 to 200 μm wide void-free zone near the crystal surface and a large void-free region in the center (Figure 11.21a). In this case, all the voids are situated in a thin layer at a distance of 50 to 200 μm from the lateral surface. As the growth rate increases, the near-surface layer remains free of voids, but the rest of the cross-section may contain voids. These voids are located primarily in planes making an angle $\leq 90°$ with the crystallization front. As the crystallization rate increases further, this mode of void distribution remains in the center of the crystal, but the void-free zones near the surface disappear (Figure 11.21b).

The above features of the void distribution in shaped sapphire crystals are not strictly connected only to the growth rate. The void distribution over the crystal cross-section is determined by the shape of the crystallization front. The latter, in turn, depends on whether or not the conditions determining stability or instability of a plane crystal-melt interface are satisfied.

In [17] the morphological stability was investigated with respect to an infinitesimal sinusoidal perturbation of a moving, planer crystal-melt interface during unidirectional crystallization of a binary alloy. Instability occurs if the amplitude of any Fourier-component of an arbitrary perturbation grows; if all the components decrease, then the interface is stable. The stability criteria derived in that paper connect the growth parameters with the system characteristics. It is shown that, in the stable state, there is a critical frequency ω_c for which the system exhibits the smallest resistance to an instability; this enables one to obtain the size of cells of an instability front:

$$\lambda = 2\pi/\omega_c = 2\pi(T_m\gamma/LG)^{1/2},$$

where T_m is the melting temperature, γ the specific interface free energy, L the heat of fusion and G the generalized temperature gradient. For real crystallization conditions of sapphire: $T_m = 2320$ K, $\gamma \approx 300$ erg/cm, $L = 3.4$ erg/cm^3, $G = 5 - 50$ K/cm, and the size of cells is approximately $40 - 150\mu$m.

In [160] an analysis was performed of the effects of surface tension anisotropy and interphase boundary kinetics on the morphological stability of a plane crystal-

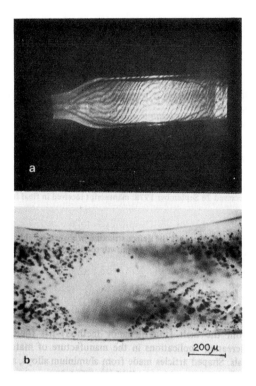

Fig. 11.20.   Tree-like void distribution in a sapphire tube grown at low rates (< 1 mm/min): (a) general view of the tube; (b) void distribution over the cros-section of the tube.

melt interface. In particular, it has been shown that the radius of curvature of the interface is, to a first approximation, the perturbation amplitude of the perturbed surface. Recent experimental evidence lends support to the above analysis.

The following explanation is proposed, therefore, for the void distribution observed in shaped sapphire crystals. At very low crystallization rates, the interphase boundary is plane and voids are not trapped. In [161] the critical growth rate has been calculated below which a foreign particle is not trapped by the grown crystal. The critical growth rate for the sapphire is approximately 0.5 mm/min.

The plane interface may become unstable in some regions; convective currents in the melt may stimulate this process. Macroscopic concavities, with a diameter of up to 1mm, arise on the crystallization front; these concavities are characterized by an enhanced trapping of voids. We suppose that the change of position of the concavities on the crystallization front with time depends on the pattern of the convective flows in the melt which leads to the void distribution shown in Figure 11.20. This process is in accordance with the Mullins and Sekerka theory for the lowest frequencies.

When growing shaped sapphire crystals at rates of 1 to 3 mm/min^{-1}, the interface

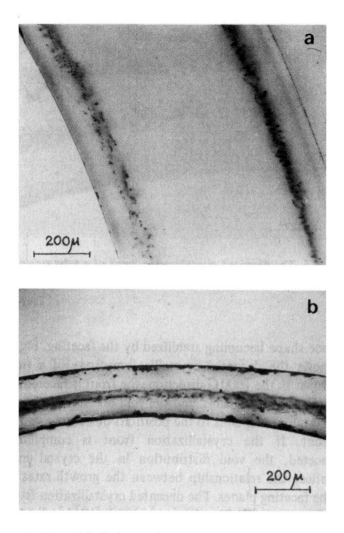

Fig. 11.21.   The void distribution over the cross-section at growth rates of 1–3 mm/min.

has, as a rule, a plane central part and convex peripheral sections. In this case, the curvature of the crystallization front plays a role of the perturbation, and conditions of instability occur primarily at regions of maximum curvature at a distance $\lambda_c = 2\pi/\omega_c$ from the interface edge, where ω_c is the critical frequency. If the crystal cross-sectional dimension $d > 2\lambda_c$, then two bands are observed in the void pattern corresponding to the regions where instability occurs. If the cross-sectional dimension is small ($d \leq 2\lambda_c$), then the interface does not contain the plane part; the instability occurs in the central part of the cross-section and the void distribution is of the type shown in Figure 11.21b. Figure 11.23 shows the decanted crystallization

Fig. 11.22. The cross-section of a tube pulled at a rate greater than 5 mm/min.

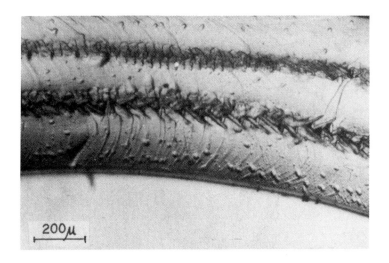

Fig. 11.23. The decanted crystallization front of a tube grown at rates of 1–3 mm/min.

front obtained at the same growth rates. The unstable region of the interface is the system of hollows and crests which is faceted by close-packed faces as the instability proceeds. Sometimes, smaller cells can be observed at the interface (Figure 11.24); these probably correspond to perturbations of higher harmonics.

As the growth rate is increased further, the entire solid-liquid interface becomes

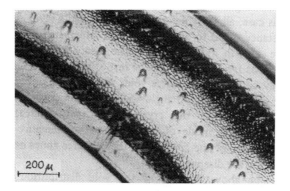

Fig. 11.24. The crystallization front of a tube grown at 1–3 mm/min showing less-commonly observed fine cell structure.

unstable. The crystallization front becomes faceted, with the new interface shape becoming stabilized by the faceting. Figure 11.25 shows the decanted crystallization front of a tube grown in the [0001] direction; the front is faceted by rhombohedral planes, and the void distribution in the crystal corresponds to the positions of hollows on the front. If the crystallization front is completely faceted, the void distribution in the crystal may reflect the relationship between the growth rates of the faceting planes. The decanted crystallization front of a tube faceted by the $\{11\bar{2}0\}$ and $\{10\bar{1}1\}$ planes is shown in Figure 11.26a. From geometrical considerations (Figure 11.26b), it is apparent that the angle γ between the crystal pulling direction and the direction of the band of void pile-up is given by $\tan\gamma = V_3/V_4$, where $\vec{V}_3 = \vec{V}_1 + \vec{V}_2$ is the vector sum of growth rates of facets 1 and 2, and V_4 is the pulling rate. Knowing the direction of the void pile-up band, one can estimate the ratio of crystallization rates of faces 1 and 2:

$$V_2/V_1 = \sin(\alpha + \gamma)/\cos(\alpha + \gamma)$$

where α is the angle between the normal to face 1 and the pulling direction. For the case of Figure 11.26, we have

$$V_2/V_1 = V_{\{11\bar{2}0\}}/V_{\{10\bar{1}1\}} = 0.965;$$

the actual growth rates of the facets can be calculated from the known pulling rate. It is noteworthy that the rhombohedral planes are most frequently found in interface faceting.

11.4.3. *Low-Angle Grain Boundaries*

The dislocation defect structure of shaped sapphire crystals was studied both by the optical polarization method and by the technique of widely diverging X-ray beams. It has been established that, in the absence of low-angle grain boundaries propagating

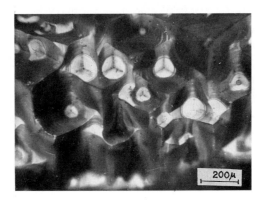

Fig. 11.25.   The crystallization front of a tube grown in the [0001] direction showing rhombohedral faceting.

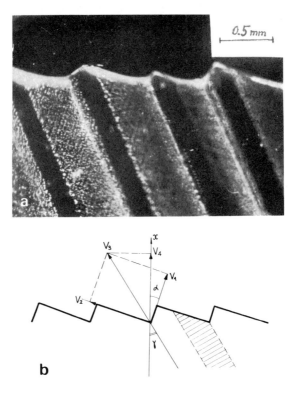

Fig. 11.26.   (a) The decanted crystallization front of a tube pulled at a rate greater than 5 mm/min. (b) Schematic of faceted crystallization front in (a) used to calculate the ratio of crystallization rates of facets.

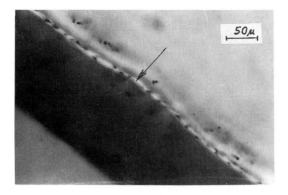

Fig. 11.27. The cross-section of a tube with a subgrain boundary passing along the void pile-up plane (shown by arrow).

from the seed, the initial part of the crystal does not contain subgrain (or low-angle grain) boundaries. Then, as the crystal grows the dislocation density increases and subgrain boundaries are formed. The misorientation of adjacent subgrains increases with distance from the seed, reaches a certain limit, and then decreases slightly with further growth. The decrease is probably associated with a rearrangement of the subgrain structure; i.e. subgrains with large misorientations branch into a series of subgrain boundaries with smaller misorientations. In addition, there is a decrease in the density of subgrain boundaries which are at large angles to the growth direction. The subgrain boundaries which remain are approximately parallel to the growth direction; the density of these boundaries (with misorientations as high as 5° to 10°) stays constant.

It should be noted that, at high crystallization rates, crystals which are free of low-angle grain boundaries can sometimes be grown. This is probably explained by the fact that the time spent by the growing crystal in the plastic zone is not long enough for polygonization processes to occur.

The presence of voids is a source of additional stresses which gives rise to dislocation generation and boundary formation. Subgrain boundaries are frequently observed to form along planes of void pile-up, as shown in Figure 11.27.

Thus, the defect structure in sapphire tube crystals is related to the distortion of the planar solid-liquid interface which, in turn, is controlled by the growth conditions.

Chapter 12

CRYSTALLIZATION UNDER MICROGRAVITY CONDITIONS

Microgravity conditions provide prerequisites for the production of materials with qualitatively new properties through melt crystallization. In particular, it is possible to obtain pseudoalloys of terrestrially immiscible components and also microhomogeneous solid solutions. The advantages of the microgravity conditions relative to these problems are being widely researched.

At the same time space technology is attracting considerable attention for the production of single crystals under microgravity conditions. The absence of gravity and of free convection (appearing in the melt due to the temperature gradients in conjunction with the gravitation) are the necessary prerequisites for producing high-quality single crystals in space.

Sufficient experience has now been gained in growing crystals in space by ampoule methods which had some important disadvantages. The ampoule itself can act as a source of melt contamination. More promising are methods for crystallization from the melt in which the lateral surface of the crystal is shaped without coming into contact with solid walls – the capillary shaping techniques. Gravity is an important factor affecting the crystallization process when the capillary shaping techniques are used. Therefore, with technical facilities to perform microgravity crystallization now available, it is important to study the gravity effect on the stability of crystallization carried out using these techniques. The above data from Chapters 2–5 for small Bond numbers can be used [121, 162].

Application of the stable growth concept to crystallization using the Verneuil technique is in particular an illustrative example. As was shown above, stable-unstable process transition is first of all determined by the capillary effects. With the meniscus height being linearly dependent on the crystal diameter (small Bond numbers), i.e., $h = R(1 + \cos \alpha_0) \sin^{-1} \alpha_0$, the process is stable. Under gravity conditions it is possible only for small diameters, i.e., for sapphire $2R \ll 6$mm.(capillary constant for sapphire). However, under microgravity conditions, e.g., on board a space station, for level of microgravity $(10^{-4} - 10^{-6})g_0$ capillary constant $a_s \approx (10^{-2} - 10^{-3})a_0$, hence, the region of stability is two to three orders larger: $2R \ll (600 \div 6000)$mm.

269

This is a very important conclusion, and is of great interest due to the works on crystal growth conducted in space. An oxygen burner should not necessarily be used as a heat source. Therefore, any heat source can be used provided the requirement obtained above (4.28) is adhered to. In addition to process stability, large-dimensional crystal growth by VT under microgravity conditions can offer additional positive effects. Firstly, the melted-layer thickness significantly increases, which decreases the temperature gradients at the crystallization front and correspondingly decreases thermal stresses. Secondly, for thick melt layers the size of the particles fed can be increased. This significantly reduces the requirements to the initial charge simplifying its preparation.

As was mentioned above, this result is in good agreement with experimental data on the growing of sapphire single crystals of small diameters by the Verneuil technique in terrestrial conditions.

12.1. Simulation Experiments

Our experiments under space conditions were preceded by simulating the liquid column shape by using immiscible liquids of equal densities. Simulation of capillary shaping under zero gravity was carried out by the following scheme [121, 163]. A column of alcohol-water solution was formed between two glass tubes, surrounded by the equal-in- density mineral oil. Pressure d in meniscus was equal to the weight of column of the alcohol solution in the upper tube. The lower tube of $2r_0$ in diameter imitated the shaper, and the upper one of $2R$ in diameter the crystal.

Parameters d, α_0 and z were determined experimentally. The results agree with the conclusion of Chapter 3 (Figure 12.1, 12.2).

12.2. Crystallization of Copper under Short-Time Microgravity Conditions

Simulation experiments with liquids cannot be used to simulate real conditions on the crystal-melt boundary. We therefore investigated the crystallization of copper under the conditions of capillary shaping on a high-altitude rocket [163,164]. The metal has a relatively low melting point $(1083^0 C)$, resistance to overloading, and its physic-chemical properties are well-known.

The technique of capillary shaping was used to produce a circular column of melt, 5–8 mm in diameter, in an evacuated enclosure. This enables us to reduce the amount of gas in the metal. Specimen 6 (Figure 12.3) was in the form of a cylinder, 5–8 mm in diameter, and about 5–6 mm long was placed between two molybdenum shapers 1 and 8 into which copper 4 was fused preliminarily. This guaranteed complete wetting of the shaper during the melting of the specimen. Two graphite guard rings 3 and 7 were introduced to prevent the escape of the melt beyond the sharp lip of the shaper. These rings were supported by the graphite spacer 5 which also acted as a thermal shield. The two shapers were pushed into the coupling tube 2. The above design

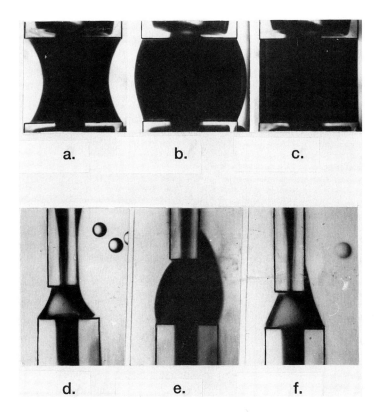

Fig. 12.1. Meniscus model appearing on pulling a circular rod (a, b, c) and plate (d, e, f) under zero-g conditions: (a, d) concave meniscus; (b, e) convex meniscus; (c) meniscus in the form of a right circular cylinder; (f) meniscus of a straight profile.

ensured rigidity, simplicity of assembly, and constant separation between the shapers. Tungsten-rhenium thermocouples were mounted near the ends of the specimens and were used to estimate the temperature distribution along its length. The capsule was inserted into a heating device which used the energy of the exothermic chemical reaction. Heat was removed from the capsule by crystallization near the upper shaper 8, the diameter of which was smaller than the diameter of the lower shaper.

For a given pressure d, a family of profile curves can be calculated. Figure 12.4 shows these profile curves for a copper melt with

$$\gamma = 1351 \text{ erg/cm}^3, \quad \rho_m = 8.21 \text{ g cm}^{-3}, \quad Z(R) = 0.5 \text{ cm}, \quad R = 0.289 \text{ cm}.$$

It is clear from the graphs that there are a number of solutions that satisfy the conditions $Z \mid_{r=r_0} = 0$ depending on the combination of α_0 and d. However, we require a unique solution which can be found by demanding that the volume of the body obtained by rotating the curve $z = f(r)$ found above. Moreover, in our

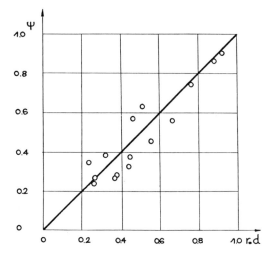

Fig. 12.2.   The theoretical $\psi = f(r_0 d)$ relation, where $\psi = (\sin \alpha_1 - R/r_0)(1 - R/r_0)^{-1}$, and experimental data.

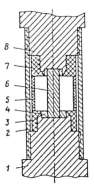

Fig. 12.3.   Design of capsule used to investigate the process of crystallization from a melt under microgravity conditions (notations are in the text).

calculation it is also important to take into account the angle of growth ψ_0, which is equal to zero in the case of copper [75].

The above scheme was used to calculate all the profile curves for columns of melt produced by capillary shaping. Assuming that the crystallization front was flat, we determined the shape of the lateral surface of the crystallized specimens.

During the experiments with high-altitude rockets, we had no photographic facilities for recording the crystallization process. Therefore the shape of the crystallized specimens was examined experimentally and was used to compare them with the calculated profiles (Figure 12.5). The dashed curves are the calculated profile of the

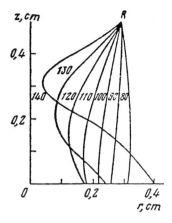

Fig. 12.4.   Calculated profile curves for a columns of a copper melt under zero-gravity conditions, $d =$ 5000 dyn.cm^{-2}. Numbers shown against curves are the values of α_0 in degrees.

melt columns and the solid curves show the calculated shape of the lateral surface of the crystallized specimen. Specimens such as 1 and 2 were crystallized under variable pressure in the copper melt whereas specimen 3 was produced under constant pressure. To achieve this, we developed a special capsule in which a column of melt, height 3 mm, could be produced under a pressure $d = 1190$ dyn cm^{-2}. Good agreement between experimental and calculated values suggests that the melt did not flow onto the crystal (this possibility was not ruled out for zero-gravity conditions – see Figure 12.1). Thermocouples mounted on the specimens enabled us to estimate the temperature distribution in the melt and the rate of crystallization, which turned out to be 5–7 mm/min^{-1}.

Our results suggested that the capillary shaping and the crucible-free zone melting could be realized in space. These crystallization processes were realized in the experiment "Shape".

12.3. "Shape" Experiment

As has been shown in 3.6.2 paragraph a cylindrical melt column ($R = r_0; \alpha_0 = \pi/2$) can be realized in the zero gravity conditions and we managed to obtain it in simulation experiments (Figure 12.1). Such a column of melt is formed under pressure d, satisfying condition $d = (2r_0)^{-1}$, but, reaching the altitude $h \approx \pi r$, it loses stability and transforms first to a meniscus, its profile curve having an ambiguous projection onto the abscissa axis, and then with increasing h, it falls into two independent meniscus.

At $h < \pi r_0$ and $d = 0a$ meniscus is a catenoid; with increasing pressure its curvature in the axial cross-section decreases, reaching infinity at $d = (2r_0)^{-1}$.

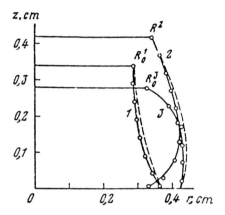

Fig. 12.5. Calculated profile curves of the initial columns of copper melt (- - - -) and crystallized specimen (———) for different pressures (1, 2, 3); experimental data of crystallized specimen shape (o o oo).

When pressure increases above $(2r_0)^{-1}$ the meniscus becomes convex.

A cylindrical meniscus can be obtained only at $\alpha_0 = \pi/2$, which means that a metal should be chosen for the material grown experimentally (for metals the angle of growth is usually close to zero).

In the conditions of microgravity at "Salyut" orbital space probe crystallization of indium was carried out using TPS [165,163]. The advantages of indium are its low melting temperature (156°C), and a comparatively high density in a solid state $(7.28g/cm^3)$ which only slightly differs from that of the melt $(7.03 gcm^3)$. The surface tension of the melt $\gamma = 592$ erg/cm^2, and capillary constant under terrestrial conditions $a = 0.41$ cm.

Figure 12.6 depicts a scheme of the growth device. A plastic case 1 with a lid 2 has a graphite container 3 filled with indium preliminary. A resistive heater 4, separated by a foam-polyurethane layer 5, is used for heating. The heat is delivered to the container through a copper capsule 6, which also serves for holding a copper cap 7 (shaper itself) and supplied with a hole for leveling the inert gas pressure inside and outside the container 3. The meniscus of melt 8 is formed first between the initial copper rod 9, fastened to a rod 10, and the edge of the shaper 7. The meniscus shape is fixed by a photo camera with an illumination system of windows 11.

Figure 12.7 depicts the shape of a drop, formed at the edge of the shaper (a), and of the meniscus of melt (b).

Pressure $d \approx 1.3$ can be found from the shape of the drop, according to the evident ratio $d = 1/R_0$ (R_0 is the radius of a drop). The product $r_0 d$ is about 0.72, that exceeds by approximately 40% that required for formation of a cylindrical column (Figure 12.7a).

After the initial rod was wetted with melt, it was pulled at the velocity $V \approx 3$ mm/min. The velocity of pulling the test indium specimen, grown by means of an analogous device in terrestrial conditions, could not be elevated above 0.2 mm/min.

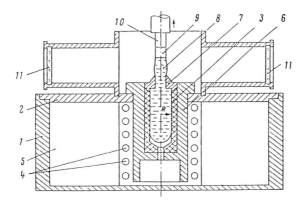

Fig. 12.6. Experimental setup for crystal growth (notations are in the text).

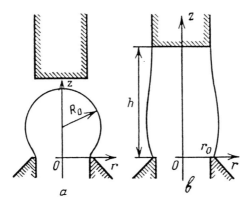

Fig. 12.7. The melt drop at the edge of the shaper (a) and the meniscus formed there upon (b).

The typical shape of a meniscus has the ratio $h/r_0 \approx 3.6 > \pi$. The meniscus height achieved in the experiment does therefore not exceed the limiting permissible one for a cylindrical melt column (Figure 12.7b).

The flight specimen diameter was approximately 5.4–5.6 mm (Figure 12.7), of the test one it was 2.9-3.2 mm. Both samples were mainly single crystals in structure.

Grown specimens were used for producing transversal micro section specimens by polishing with a soap solution of aluminum oxide, an alcohol solution of picrine acid and further chemical etching by an aqueous solution of indium trichloride. The material of the sample contains a great amount of gaseous inclusions of various shape and dimension, whereas no accluded gas was observed in the test specimen. A micro analysis of thin sections showed the presence of copper up to 1%.

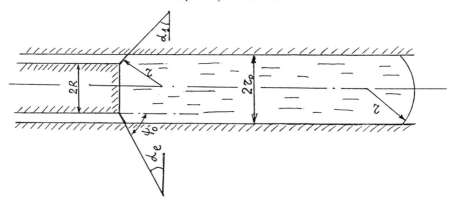

Fig. 12.8.   The schematic drawing of tape-growth space experiment.

12.4. "Ribbon" Experiment

The crystallization of ribbons from the melt of Ge and GaAs in the orbital station "Mir" is the next example of the capillary shaping technique at microgravity conditions (Figure 12.8) [166]. The melt is placed in a fixed gap between two flat walls. In practice it is the realization of TPS with the boundary conditions of wetting on the walls of the shaper. The employment of unwetted walls was successful, if the condition: $\alpha_1 > \alpha_0$ that is $\theta > \pi - \psi_0$ is satisfied.

The meniscus projection is part of a circle and the dependence between the ribbon thickness $2R$ and the gap $2r_0$ is:

$$R = r_0 \frac{\sin \alpha_0}{\sin \alpha_1}. \tag{12.1}$$

The capillary coefficients of set (2.1) are:

$$A_{RR} = -V \frac{\sin \alpha_1}{r_0 \cos \alpha_0} \tag{12.2}$$

$$A_{Rh} = -V \frac{\sin \alpha_1}{r_0 \sin \alpha_0}. \tag{12.3}$$

The coefficient A_{RR} is negative and there is capillary stability from a theoretical point of view.

At the flight experiments Ge and GaAs ribbons were grown between walls made from pirocarbon (Figure 12.9). Unfortunately the regime of melting was not optimal, the seeds melted and the ribbons had a policrystal structure. But flat ribbon growth without special regulation of the process is evidence of process stability.

12.5. Conclusion

The tentative theoretical analysis of the crystallization process stability made by the methods of capillary shaping showed that in some cases the crystallization process

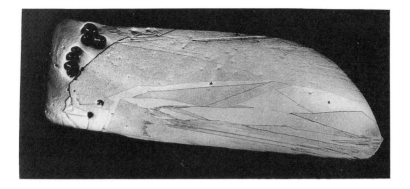

Fig. 12.9. The view of flight specimen.

is more stable in the conditions of microgravity. Model experiments on capillary shaping, carried out with two immiscible liquids, on crystallization of copper in the conditions of short-term zero gravity, on crystallization of indium in the orbital station "Salut" and on crystallization of germanium and gallium arsenide on the orbital station "Mir" showed that the process of crystallization can satisfactorily be described in the framework of the theoretical model developed in this monograph. It therefore seems reasonable that cosmic experiments should be continued.

APPENDIX: CALCULATION OF THE GROWTH ANGLE FROM THE SHAPE OF A CRYSTALLIZATION DROP

To find the equation of the generating line of a crystallized drop body of revolution the origin of the coordinate system should be placed in the point of drop crystallization initiation O_1 and turned through by an angle ϵ_0 so that the r_1-axis becomes a line tangent to the solidified drop surface (Figure 1.5). For the new coordinate system $r_1 O_1 z_1$ the profile curve equation for the crystallized drop will take the following form

$$\frac{dz_i}{dr_1} = \mathrm{tg}\psi_i \approx \psi_i, \tag{A.1}$$

where $\psi_i = \epsilon_0 - \psi_S$.

For small-values ψ_i expansion of the right side of Equation (A.1) into a Taylor series holds. If the series is restricted to the linear terms alone it follows that:

$$\frac{dz_1}{dr_1} = -\frac{d\psi_i}{dr_i} \Big|_{r_1=0} r_1. \tag{A.2}$$

Integration of (A 1.2) gives the generating-line equation that is a parabola one:

$$z_1 = K_1 r_1^2, \tag{A.3}$$

where

$$K_1 = -\frac{1}{2}\frac{d\psi_i}{dr_i} \Big|_{r_1=0}.$$

Now the solution of the set of equations:

$$z = z_1 \cos \epsilon_0 - r_1 \sin \epsilon_0$$

$$r = z_1 \sin \epsilon_0 + r_1 \cos \epsilon_0 + r_0$$

$$z_1 = K_1 r_1^2 \tag{A.4}$$

gives the profile curve equation in the initial coordinate system:

$$r = K_1 \sin \epsilon_0 \left(\frac{0.5 \text{tg} \epsilon_0}{K_1} - B \right) + \cos \epsilon_0 \left(\frac{0.5 \text{tg} \epsilon_0}{K_1} - B \right) + r_0, \qquad (A.5)$$

where

$$B = \left[\left(\frac{0.5 \text{tg} \epsilon_0}{K_1} - B \right)^2 - \frac{z}{K_1 \cos \epsilon_0} \right]^{1/2}.$$

The parameters K_1 and ϵ_0 are defined from photographs of solidified drops. When defining the angle ϵ_0 it is convenient to make use of the following property of a parabola: its portion formed by a line tangent to the parabola from the point of tangency to the point of its intersection with the z_1-axis is divided into equal parts by the r_1-axis. The validity of the approximation found is confirmed by very good agreement between the solidified drop profiles calculated from (A.5) and the real ones (the calculated points lie on the experimental drop contour)

The value of any profile point is calculated by differentiation of Equation (A.5):

$$\psi_S(r) = \text{arctg}(\text{ctg} \epsilon_0 - A^{-1}), \qquad (A.6)$$

where

$$A = [\cos^2 \epsilon_0 - 4K_1(r_0 - r) \sin \epsilon_0]^{1/2}.$$

To define the angle $\psi_L(r)$ one can make use of the fact that in the process of crystallization the drop is so small that it can be regarded as part of a sphere. The smaller the Bond number as compared with *one*, i.e., the smaller the drop base dimension as compared with the melt capillary constant, the more accurately this condition will hold. For InSb, e.g., $\gamma_{LG} = 0.425$ N m^{-1}, $\rho_L = 6480$ kg m^{-3}, $\rho_S = 5750$ kg m^{-3} [25], for the capillary constant the value of $a = 3.7 \ 10^{-3}$m is obtained, for Ge $\gamma_{LG} = 0.616$ N m^{-1}, $\rho_S = 5260$ kg m^{-3}, $\rho_L = 5510$ kg m^{-3} [115, 163] and $a = 4.8 \ 10^{-3}$m, for Si $\gamma_{LG} = 0.720$ N m^{-1}, $\rho_S = 2300$ kg m^{-3}, $\rho_L = 2530$ kg m^{-3} [163, 32] and hence $a = 7.6 \ 10^{-3}$m. When measuring linear dimensions the maximum error did not exceed 0.05 mm, the errors of ϵ_0-measurement and K_1-calculation do not exceed $\Delta \epsilon_0 < 1.5x$ and $\delta K_1 < 1\%$, respectively.

As long as the volume of the liquid drop does not exceed the volume of a hemisphere (i.e., $\psi_L \leq \pi/2$) the angle $\psi_L(r)$ is calculated from the formula (Figure 1.5a):

$$\psi_L(r) = \text{arctg}[2hr/(r^2 - h^2)], \qquad (A.7)$$

where from simple geometrical relations the following equation for the drop height h follows:

$$h^3 + 3r^2h - 6W/\pi = 0. \qquad (A.8)$$

Here W is the volume of the uncrystallized drop

$$W = \pi \rho_S \rho_L^{-1} \int_0^{z_0} r^2(z)dz - \int_0^{z} r^2(z)dz$$

z_0 is the height of the crystallized drop (Figure 1.5a). In case $\psi_L(r) > \pi/2$, from Figure 1.5b it follows that $\psi_L = \pi - \psi^*$. For ψ^* (A.7) holds and h can be calculated from the following equation:

$$4\pi R^3/3 - W = \pi h(3r^2 + h^2)/6 \tag{A.9}$$

that can be reduced to the following form by substituting $R = (r^2 + h^2)/2h$:

$$h^3 - \pi r^4 h^2/2W - \pi r^6/6W = 0 \tag{A.10}$$

Here R is the radius of the sphere, the drop being a part thereof. The growth angle was calculated from (A.5)–(A.10) using a computer and the Kardan substitution for the cubic equation (Figure 1.6).

REFERENCES

1. V. A. Tatarchenko, 1973, *Fiz. & Khim. Obrabotki Mater.* 6, 47. [Sov. Phys. & Chem. Mater. Treatment].
2. V. A. Tatarchenko, 1974, 4th International Conference on Crystal Growth, Thesis of Reports, Tokyo, pp. 521–522.
3. T. Surek and B. Chalmers, (1975), *J. Cryst. Growth* 29, 1.
4. V. A. Tatarchenko, 1988, *Stable Crystal Growth* (in Russian: Ustoichiviy Rost Kristallov), Nauka, Moscow p. 240.
5. J. Czochralski, 1917, *Zs. Phys. Chem.* 92, 219.
6. A. V. Stepanov, 1959, *Zh. Tech. Fiz.* 29, 339. [Transl. Soviet Phys. -JETP].
7. B. Chalmers, H. E. LaBell, and A. J. Mlavsky, 1972, *J. Cryst. Growth* 13/14, 84.
8. G. H. Schwuttke, 1977, *Phys. Status Solidi (a)* 43, 43.
9. M. A. Verneuil, 1904, *Annales de Chemie et de Physique, Nuitieme serie* 3, 20.
10. W. G. Pfann, 1966, *Zone melting*, J. Wiley and Sons, New York and London.
11. D. T. J. Hurle, 1977, *J. Cryst. Growth* 42, 473.
12. D. T. J. Hurle, G. C. Joyce, G. C. Wilson *et al*, *J. Cryst. Growth* 74, 480.
13. D. T. J. Hurle, G. C. Joyce, M. Ghassempoory *et al*, 1990, *J. Cryst. Growth* 100, 11.
14. V. S. Leibovich, 1983 *Bulletin of the Academy of Sciences of the USSR, Physical series* 47, 9.
15. D. E. Temkin, 1961, *Soviet Physics Doklady*. A Translation of the Physics section of the Proceedings of the USSR Academy of Sciences 5, 609.
16. V. V. Voronkov, 1964, *Fizika Tverdogo Tela* 6, 2984. [Sov. Physics, Solid State].
17. V. W. Mullins and F. R. Sekerka, 1964, *J. Appl. Phys.* 35, 444.
18. B. L. Timan and O. D. Kolotiy. In: Proceedings of the 9th Conference on Stepanov's Growth of Single Crystals (in Russian, Ioffe Phys.-Tech. Inst., Leningrad, 1982) p. 66.
19. P. Jorgenson, 1969. *Rev. Internat. Hautes Temp rat. et Refract* 6, 199.
20. V. A. Tatarchenko, 1979, *Kristallografia* 24, 238. [Sov. Phys. Crystallography].
21. V. A. Tatarchenko and L. M. Umarov, 1980, *Kristallografia* 25, 1311.
22. L. M. Umarov and V. A. Tatarchenko, 1984, *Kristallografia* 29, 1146.
23. E. V. Gomperz, 1922, *Zeitschr. für Phys.* 8, 184.
24. P. Kapitza, 1928, *Proc. Roy. Soc. Ser. A.* 119, 358.
25. Bull. Acad. Sci. USSR, Phys. Ser. (Cons. Bureau, New York): 3 (1969) No. 12, 35 (1971) No. 3, 36 (1972) No. 3, 37 (1973) No. 11, 40 (1976) No. 7, 44 (1980) No. 2, 47 (1983) No. 2 etc.
26. H. E. LaBelle, 1975, US Patent No. 3915662.
27. T. Toshiba, 1976, *J. Electron. Eng.* 5, 46.
28. S. K. Brantov, B. M. Epelbaum, and V. A. Tatarchenko, 1984, *Materials Letters* 2, 274.
29. V. A. Tatarchenko, A. I. Saet, and A. V. Stepanov, 1968, *Proceedings of the 1st Conference on Stepanov's Growth of Semiconductor Single Crystals* (in Russian, Ioffe Phys.-Tech. Inst.,

Leningrad), p. 83.

30. V. A. Tatarchenko, A. I. Saet, and A. V. Stepanov, 1969. In: Ref. [25] 33, 1782.
31. V. A. Tatarchenko, 1977 *J. Cryst. Growth* 37, 272.
32. G. V. Sachkov, V. A. Tatarchenko, and D. I. Levinzon, 1973. In: Ref. [25] 37, 2288.
33. T. Arizumi and N. Kabayashi, 1972, *J. Cryst. Growth* 13/14, 615.
34. V. A. Borodin, E. A. Brener, and V. A. Tatarchenko, 1982, *Crys. Res. and Technol.* 17, 1187.
35. V. A. Borodin, E. A. Brener, T. A. Steriopolo, V. A. Tatarchenko, and L. I. Chernishova, 1982, *Crys. Res. & Technology* 17, 1199.
36. E. A. Brener, V. A. Tatarchenko, and B. E. Fradkov, 1982, *Kristallografia* 27, 205 (Sov. Crystallography).
37. R. S. Wagner and W. C. Ellis, 1964, *Appl. Phys. Letters* 4, 89.
38. R. S. Wagner and W. C. Ellis, 1965, *Trans. Met. Soc. AIME* 233, 1053.
39. G. A. Arzumanyan, 1990, *J. Cryst. Growth* 99, 859.
40. J. D. Zook, B. G. Hoepke, B. L. Grung, and M. H. Leipold, 1980, *J. Cryst. Growth* 50, 260.
41. S. K. Brantov, V. A. Tatarchenko, and B. M. Epelbaum, 1985, *Poverchnost. Fisica, Chimia, Mechanica* 2, 139 (Sov. Surface. Phys. Chem. Mech.).
42. S. K. Brantov, B. M. Epelbaum, and V. A. Tatarchenko, 1987, *J. Cryst. Growth* 82, 122.
43. A. Mlavsky and N. A. Panidscic, 1975, US Patent No. 3868228.
44. V. A. Borodin, T. A. Steriopolo, V. A. Tatarchenko, and T. N. Jalovets, 1983. In: Ref. [25] 47, 151.
45. V. A. Borodin, T. A. Steriopolo, and V. A. Tatarchenko, 1985, *Cryst. Res. & Technol.* 20, 833.
46. V. A. Borodin, V. V. Sidorov, T. A. Steriopolo, and V. A. Tatarchenko, 1987, *J. Cryst. Growth* 82, 89.
47. P. I. Antonov, J. G. Nosov, and S. P. Nikanorov, 1985. In: Ref. [25] 49, 6.
48. I. V. Alyab'ev, S. V. Artemov, N. I. Bletzkan, and V. S. Papkov, 1985. In Ref. [25] 49, 9.
49. V. A. Borodin, V. V. Sidorov, T. A. Steriopolo, V. A. Tatarchenko, and T. N. Jalovets, 1988. In Ref. [25] 52, 2009.
50. V. A. Borodin, V. V. Sidorov, S. N. Rossolenko, T. A. Steriopolo, V. A. Tatarchenko, and T. N. Jalovets, 1990 *J. Cryst. Growth* 104, 69.
51. V. A. Tatarchenko and E. A. Brener, 1976. In: Ref. [25] 40, 1456.
52. E. A. Brener and V. A. Tatarchenko, 1978, *Acta Physica Scientiarum Hungaricae* 47, 133.
53. V. A. Tatarchenko, 1979, Rapporto Interno del Laboratorio MASPEC del Consiglio Nazionale delle Ricerche N36 (Parma, Italia).
54. V. A. Tatarchenko and E. A. Brener, 1980, *J. Cryst. Growth* 50, 33.
55. E. A. Brener and V. A. Tatarchenko, 1979. In: Ref. [25] 43, 1926.
56. V. A. Tatarchenko, 1980, *Growth of Crystals* 13, 160. (Consultants Bureau, New York).
57. E. A. Brener and V. A. Tatarchenko, 1983, *Growth of Crystals* 14, 139.
58. V. A. Tatarchenko, 1980. In: Abstracts of 4th International Specialists School on Crystal Growth (Suzdal, USSR), p. 203.
59. V. A. Tatarchenko and E. A. Brener, 1982, *Proceedings of European Meeting on Crystal Growth '82 Materials for Electronics*, Prague, p. 86.
60. G. A. Korn and T. M. Korn, 1961, *Mathematical Handbook for Scientists and Engineers*, McGraw-Hill Book Company, New York, San Francisco, Toronto, London, Sydney), p. 282.
61. V. A. Tatarchenko, 1987, *J. Cryst. Growth* 82, 74.
62. V. A. Likov, 1978, *Handbook on Heat and Mass Transfer*, Energia, Moscow, p. 132.
63. L. D. Landau and E. M. Lifchits, 1971, *Mecanique des Fluids*, Mir, Moscow, p. 289.
64. R. Courant and D. Hilbert, 1953, *Methods of Mathematical Physics*, Interscience Publishers, New York, vol. 1, p. 169.
65. L. S. Srubschik and V. I. Yudovich, 1966, *USSR Computational Mathematics and Mathematical Physics (Oxford)* 6, 1127.
66. P. Funk, 1970, *Variationsrechnung und Anwendung in Physik und Technik*, Springer, Berlin, Heidelberg, New York, p. 132.
67. G. K. Gaule and J. R. Pastore, 1961, 'The role of surface tension in pulling single crystals of controlled dimensions'. In: *Metallurgy of Elemental and Compound Semiconductors*, Interscience, New York, London, p. 201.
68. B. M. Goltsman, 1961, *Growth of Crystals* 3, 408 (Consultants Bureau, New York).

69. Ju. M. Shashkov and E. V. Melnikov, 1965, *Zh. Fiz. Khim.* 39, 1364 (Sov. J. Phys. Chim.).
70. P. I. Antonov, 1965, *Growth of Crystals* 6, 158 (Consultants Bureau, New York).
71. S. V. Tsivinskiy, 1962, *Inzh. Fiz. Zh.* 5, 59 (Sov. J. Eng. Phys.).
72. A. I. Pogodin, I. M. Tumin, and A. M. Eidenzon, 1973. In: Ref. [25] 37, 2292.
73. V. V. Voronkov, 1974, *Kristallografia* 19, 228 (Sov. Crystallography).
74. Yu. V. Naidich, N. F. Grigorenko, and V. M. Perevertailo, 1980. In: Ref. [25] 44, 236.
75. H. Wenzel, A. Fattah, and W. Uelhoff, 1976, *J. Cryst. Growth* 36, 319.
76. G. A. Satunkin, V. A. Tatarchenko, and V. Y. Shaitanov, 1980, *J. Cryst. Growth* 50, 133.
77. V. V. Voronkov, 1963, *Fiz. Tv. Tela* 5, 571 (Sov. Solid St. Phys.).
78. V. V. Voronkov, 1972, *Kristallografia* 17, 909 (Sov. Crystallography).
79. V. V. Voronkov, 1974, *Kristallografia* 19, 922.
80. V. V. Voronkov, 1978, *Kristallografia* 23, 249.
81. W. Bardsley, F. C. Frank, G. W. Green, and D. T. J. Hurle, 1974, *J. Crystal Growth* 23, 341.
82. G. F. Bolling and W. A. Tiller, 1960, *J. Appl. Phys.* 31, 1345.
83. B. Chalmers, 1964, *Principles of Solidification*, John Wiley and Sons, New York, London, Sydney.
84. Yu. F. Schelkin, 1971, *Fizika & Chimia Obrab. Mater.* 3, 29 (Sov. Phys. and Chem. Mater. Treatm.).
85. S. V. Tsivinskiy and A. V. Stepanov, 1974, *Physics of Condenced State* 33, 82 (In Russ. Trans. of Low Temp. Inst., Kharkov).
86. C. Herring, 1951, *The Physics of Powder Metallurgy*, McGraw Hill Co., New York, Paris, p. 143.
87. H. S. Carslaw and J. C. Jaeger, 1959, *Conduction of Heat in Solids*, Clarendon Press, Oxford, p. 148.
88. E. Billig, 1956, *Proc. Roy. Soc. 235A*, N1200, 37.
89. V. V. Solomatov, 1965, *Sov. Phys. Journ.* 5, 86 (New York, Transl. of Izvestija Vysshikh Uchebnykh Zavedenij, Fizika).
90. B. M. Goltsman, 1958, *Sov. Phys. Journ.* 6, 130.
91. V. A. Tatarchenko, 1976, *J. Eng. Phys.* 30, 532 (New York, Transl. of Inzhenerno-Fizicheskij Zhurnal).
92. D. Schwabe, A. Scharmen, F. Presser, and R. Ocder, 1978, *J. Cryst. Growth* 43, 305.
93. R. Pohl, 1954, *J. Appl. Phys.* 23, 668.
94. K. Mika and W. Uelhoff, 1975, *J. Cryst. Growth* 30, 9.
95. M. M. Nicolson, 1949, *Proc. Cambridge Phyl. Soc.* 45, 288.
96. W. Uelhoff and K. Mika, 1975. In: *Berichte der Kernforschungsanlage Jülich* 2c, 1195.
97. V. V. Voronkov, 1963, *Nauchn. Trudi Giredmeta* 10, 251 [In Russ. Proc. Inst. Rare Metals].
98. D. F. Games, 1974, *J. Fluid Mech.* 63, 657.
99. A. Ferguson, 1912, *Phil. Mag. Series 6* 24, 837.
100. W. Heywang, 1956, *Z. Naturforschg.* 11a, 238.
101. E. A. Boucher and T. G. I. Jones, 1980, J. C. S. Faraday 1 V. 76, Part 11, 1419.
102. G. Bakker, 1928, *Handbuch der Experimental-Physik*, Vol. 6, Leipzig.
103. G. A. Poincaret, 1985, *Course de Physique Mathem. Capillarite*, Paris, p. 163.
104. C. Huh and L. E. Scriven, 1969, *J. Colloid Interface Sci.* 30, 323.
105. Boshforth Adams, 1883, *An Attempt to Test the Theories of Capillary Action*, Cambridge.
106. S. Sugden, 1921, *J. Chem. Soc.*, 1483.
107. N. K. Adam, 1941, *The Physics and Chemistry of Surfaces*, Cambridge.
108. J. C. Brice, 1968, *J. Cryst. Growth* 2, 395.
109. W. R. Wilcox and R. L. Duty, 1966, *J. Heat Transfer. Trans. AIME* 88C, 45.
110. A. Van der Hart and W. Uelhoff, 1981, *J. Cryst. Growth* 51, 251.
111. L. Buckley-Golder and C. J. Humprey, 1979, *Phil. Mag.* A39, 41.
112. B. L. Timan and O. D. Kolotiy, 1979. In: *Thesis of Report of 1st Allunion Conference "Perspectives of Development of Techniques of Crystal Growth"*, Charkov p. 67 [in Russ.].
113. K. J. Bachman, H. J. Kirsch, and K. J. Vetter, 1970, *J. Cryst. Growth* 7, 290.
114. V. M. Goldfarb, A. V. Donskoy, and A. V. Stepanov, 1965, *Proceedings of Leningrad Pedagogic Institute* 265, 61 [in Russ.].
115. S. V. Tsivinski, P. I. Antonov, and A. V. Stepanov, 1970, *Zh. Tekh. Fiz.* 40, 372 [Transl. Soviet Phys. - Techn. Phys.].
116. A. I. Saet and V. A. Tatarchenko, 1969, *Zh. Tekh. Fiz.* 39, 1302.

117. P. F. Byrd, 1954, *Elliptik Integrals for Engineers and Physisists*, Berlin.
118. V. A. Tatarchenko and E. A. Brener, 1980, 'Influence of Surface Tension Gradient on Meniscus Processes at Crystal Growth from Melt', in: *Materials and Processes of Space Technology*, Nauka, Moscow, p. 75 [in Russ.].
119. V. A. Tatarchenko, B. N. Korchunov, N. A. Gunko, and A. V. Stepanov, 1972. In Ref. [25] 36, 476.
120. V. A. Tatarchenko, D. I. Levinzon and A. V. Stepanov, 1972. In Ref. [25] 36, 471.
121. V. A. Tatarchenko and S. K. Brantov, 1976. In Ref. [25] 40, 116.
122. G. A. Satunkin and V. A. Tatarchenko, 1985, *J. of Colloid and Interface Science* 104, 318.
123. E. A. Brener and V. A. Tatarchenko, 1983. In: Ref. [25] 47 N2, 25.
124. V. A. Tatarchenko, 1983. In Ref. [25] 47 N2, 20.
125. V. A. Tatarchenko and G. A. Satunkin, 1976. In Ref. [25] 40, 1488.
126. V. A. Tatarchenko and G. A. Satunkin, 1977, *J. Cryst. Growth* 37, 285.
127. V. A. Tatarchenko, V. V. Vachrushev, A. S. Kostygov, and A. V. Stepanov, 1971. In Ref. [25] 35, 511.
128. Sh. O. Arzumanian, L. A. Litvinov, and S. N. Shorin, 1973, *Monokristalli i Technika* 2, 201 [Sov. Single Crystals & Technique, Kharkov].
129. V. A. Tatarchenko and G. I. Romanova, 1973, *Monokristalli i Technika* 2, 48 [Sov. Single Crystals & Technique, Kharkov].
130. G. I. Romanova, V. A. Tatarchenko, and N. P. Tichonova, 1976. In: Trudi GOI, V. 54, N188, p. 10 (Proceedings of State Optical Institute, Leningrad, in Russ.).
131. V. A. Borodin, E. A. Brener, V. A. Tatarchenko *et al*, 1981, *J. Cryst. Growth* 52. 505.
132. T. Surek and S. R. Coriell, 1977, *J. Cryst. Growth* 37, 253.
133. E. A. Brener, G. A. Satunkin, and V. A. Tatarchenko, 1978, *Acta Physica Scientiarum Hungaricae* 47, 159.
134. E. A. Brener, G. A. Satunkin, and V. A. Tatarchenko, 1983, *Growth of Crystals* 14, 153 (Consult. Bureau, New York).
135. E. I. Givargizov, 1977, *The Growth of Whiskers and Plate Crystals from Vapor* (in Russian), Nauka, Moscow.
136. S. A. Ammer and V. B. Fursov, 1980, *Fiz. i Khim. Obr. Mater.* 4, 71 [Sov. Phys. & Chem. Mater. Treatm.).
137. G. A. Satunkin and V. A. Tatarchenko, 1985, *Kristallografia* 30, 772 [Sov. Crystallography].
138. E. I. Givargizov and A. A. Chernov, 1973, *Kristallografia* 18, 147 (Sov. Crystallography).
139. J. E. Geguzin and A. S. Dzuba, 1981, *J. Cryst. Growth* 52, 337.
140. L. D. Landau and V. G. Levich, 1942, *Acta Phys. Chim. URSS* 17, 42.
141. G. I. Babkin, E. A. Brener, and V. A. Tatarchenko, 1980, *J. Cryst. Growth* 50, 45.
142. J. T. H. Pollock, 1972, *J. Mater. Science* 7, 632.
143. B. L. Timan and O. D. Kolotiy, 1980, *Rost i Svoistva Kristallov* 6, 134 [Sov. Growth & Properties of Crystals, Charkov).
144. J. A. Burton, P. C. Prim, and W. P. Slichter, 1953, *J. Chem. Phys.* 21, 1987.
145. J. C. Swartz, T. Surek, and B. Chalmers, 1975, *J. Electron. Mater.* 4, 255.
146. S. K. Brantov and V. A. Tatarchenko, 1983, *Crys. Res. & Techn.* 18, K59.
147. J. P. Kalejes, 1978, *J. Cryst. Growth* 44, 329.
148. V. A. Tatarchenko, S. K. Brantov, and N. V. Abrosimov, 1978, *Fiz. & Khim. Obrab. Mater.* 1, 79 (Sov. Phys. & Chem. Mater. Treatm.).
149. N. V. Abrosimov, S. K. Brantov, B. Ljuks, and V. A. Tatarchenko, 1983. In Ref. [32] 47, 351.
150. Yu. A. Osip'yan and V. A. Tatarchenko, 1983. In Ref. [25] 47, 346.
151. J. Shi, G. Geitos, N. V. Abrosimov, S. K. Brantov, S. A. Erofeeva, and V. A. Tatarchenko, 1979. In Ref. [25] 43, 174.
152. N. V. Abrosimov, A. V. Bazhenov, and V. A. Tatarchenko, 1987, *J. Cryst. Growth* 82, 203.
153. N. V. Abrosimov, A. V. Bazhenov, V. A. Goncharov, and S. A. Erofeeva, 1983. In Ref. [32] 47, 134.
154. M. N. Leipold, 1977, *Proc. Photovolt. Solar Energy Conf.*, Luxemburg, p. 872.
155. N. V. Abrosimov, S. K. Brantov, V. A. Tatarchenko, and B. M. Epelbaum, 1982, Inorganic Mater. 28, 181 (Cons. Bureau, New York).
156. B. S. Red'kin, V. N. Kurlov and V. A. Tatarchenko, 1987, *J. Cryst. Growth* 82, 106.
157. K. F. Eskin, B. S. Red'kin, V. A. Tatarchenko *et al*, 1987, *Soviet Phys.-Techn. Phys.* 32, 1863

[Transl. of Zhurnal Tekhnicheskoi Fiziki).

158. B. S. Red'kin, V. N. Kurlov, and V. A. Tatarchenko, 1985. In Ref. [25] 49, 2412.
159. V. A. Tatarchenko, T. N. Jalovets, G. A. Satunkin *et al*, 1980, *J. Cryst. Growth* 50, 335.
160. S. Coriell and R. Sekerka, 1976, *J. Cryst. Growth* 34, 157.
161. A. A. Chernov, D. T. Temkin, and A. M. Melnikova, 1976, *Kristallografia* 21, 652 (Sov. Crystallography).
162. V. A. Tatarchenko, 1991, *Adv. Space Res.* 11, 307.
163. Ju. A. Osip'ian and V. A. Tatarchenko, 1988, *Adv. Space Res.* 8, (12)17.
164. M. S. Agafonov, G. A. Gavrilov, L. N. Gubina, V. A. Tatarchenko *et al*, 1979. In Ref. [25] 43, 124.
165. V. A. Tatarchenko, S. K. Brantov, V. L. Levtov *et al*, 1985. In Ref. [25] 49, 77.
166. E. V. Markov and V. A. Tatarchenko, to be published.

Mechanics

FLUID MECHANICS AND ITS APPLICATIONS

Series Editor: R. Moreau

Aims and Scope of the Series

The purpose of this series is to focus on subjects in which fluid mechanics plays a fundamental role. As well as the more traditional applications of aeronautics, hydraulics, heat and mass transfer etc., books will be published dealing with topics which are currently in a state of rapid development, such as turbulence, suspensions and multiphase fluids, super and hypersonic flows and numerical modelling techniques. It is a widely held view that it is the interdisciplinary subjects that will receive intense scientific attention, bringing them to the forefront of technological advancement. Fluids have the ability to transport matter and its properties as well as transmit force, therefore fluid mechanics is a subject that is particularly open to cross fertilisation with other sciences and disciplines of engineering. The subject of fluid mechanics will be highly relevant in domains such as chemical, metallurgical, biological and ecological engineering. This series is particularly open to such new multidisciplinary domains.

1. M. Lesieur: *Turbulence in Fluids.* 2nd rev. ed., 1990   ISBN 0-7923-0645-7
2. O. Métais and M. Lesieur (eds.): *Turbulence and Coherent Structures.* 1991
      ISBN 0-7923-0646-5
3. R. Moreau: *Magnetohydrodynamics.* 1990   ISBN 0-7923-0937-5
4. E. Coustols (ed.): *Turbulence Control by Passive Means.* 1990   ISBN 0-7923-1020-9
5. A.A. Borissov (ed.): *Dynamic Structure of Detonation in Gaseous and Dispersed Media.* 1991
      ISBN 0-7923-1340-2
6. K.-S. Choi (ed.): *Recent Developments in Turbulence Management.* 1991
      ISBN 0-7923-1477-8
7. E.P. Evans and B. Coulbeck (eds.): *Pipeline Systems.* 1992   ISBN 0-7923-1668-1
8. B. Nau (ed.): *Fluid Sealing.* 1992   ISBN 0-7923-1669-X
9. T.K.S. Murthy (ed.): *Computational Methods in Hypersonic Aerodynamics.* 1992
      ISBN 0-7923-1673-8
10. R. King (ed.): *Fluid Mechanics of Mixing.* Modelling, Operations and Experimental Techniques. 1992   ISBN 0-7923-1720-3
11. Z. Han and X. Yin: *Shock Dynamics.* 1993   ISBN 0-7923-1746-7
12. L. Svarovsky and M.T. Thew (eds.): *Hydroclones.* Analysis and Applications. 1992
      ISBN 0-7923-1876-5
13. A. Lichtarowicz (ed.): *Jet Cutting Technology.* 1992   ISBN 0-7923-1979-6
14. F.T.M. Nieuwstadt (ed.): *Flow Visualization and Image Analysis.* 1993   ISBN 0-7923-1994-X
15. A.J. Saul (ed.): *Floods and Flood Management.* 1992   ISBN 0-7923-2078-6
16. D.E. Ashpis, T.B. Gatski and R. Hirsh (eds.): *Instabilities and Turbulence in Engineering Flows.* 1993   ISBN 0-7923-2161-8
17. R.S. Azad: *The Atmospheric Boundary Layer for Engineers.* 1993   ISBN 0-7923-2187-1
18. F.T.M. Nieuwstadt (ed.): *Advances in Turbulence IV.* 1993   ISBN 0-7923-2282-7
19. K.K. Prasad (ed.): *Further Developments in Turbulence Management.* 1993
      ISBN 0-7923-2291-6
20. Y.A. Tatarchenko: *Shaped Crystal Growth.* 1993   ISBN 0-7923-2419-6

Kluwer Academic Publishers – Dordrecht / Boston / London

Mechanics

FLUID MECHANICS AND ITS APPLICATIONS
Series Editor: R. Moreau

21. J.P. Bonnet and M.N. Glauser (eds.): *Eddy Structure Identification is Free Turbulent Shear Flows.* 1993   ISBN 0-7923-2449-8

Kluwer Academic Publishers – Dordrecht / Boston / London